MISSION PENGUIN

MISSION PENGUIN

A PHOTOGRAPHIC QUEST FROM THE
GALÁPAGOS TO ANTARCTICA

URSULA CLARE FRANKLIN

BLOOMSBURY WILDLIFE

LONDON · OXFORD · NEW YORK · NEW DELHI · SYDNEY

BLOOMSBURY WILDLIFE
Bloomsbury Publishing Plc
50 Bedford Square, London, WC1B 3DP, UK
29 Earlsfort Terrace, Dublin 2, Ireland

BLOOMSBURY, BLOOMSBURY WILDLIFE and the Diana logo
are trademarks of Bloomsbury Publishing Plc

First published in the United Kingdom 2024

A catalogue record for this book is available from the British Library.

Library of Congress Cataloguing-in-Publication data has been applied for.

ISBN: HB: 978-1-3994-0467-9
ePub: 978-1-3994-0468-6
ePDF: 978-1-3994-0469-3

10 9 8 7 6 5 4 3 2 1

Designed by Austin Taylor
Printed and bound in China by RR Donnelley Asia Printing Solutions Ltd,
Dongguan Guangdong

MIX
Paper | Supporting
responsible forestry
FSC
www.fsc.org
FSC® C144853

To find out more about our authors and books visit
www.bloomsbury.com and sign up for our newsletters.

For my friends, human and feathered

CONTENTS

PAGE 1 Two Gentoo Penguins in perfect step with each other. They almost looked as though they were dancing.

PAGE 2 Adélie Penguins crossing the sea ice of the Weddell Sea.

PREVIOUS PAGE King Penguins arriving back at Volunteer Point, East Falkland.

THESE PAGES Adélie Penguins, Antarctica.

PREFACE

In July 2012, as the rest of the world eagerly awaited the start of the London Olympics, my own world fell apart. My darling husband, Ralph, died in my arms following a major heart attack, just a few months short of our 25th wedding anniversary. We had been together for over half of my life and I was absolutely devastated.

Ralph's death was my third close bereavement in just 10 years as I had already lost my beautiful eldest sister and my inspirational mother. Through these experiences I had learned what I needed to do to cope and had become more resilient. Grief is a very personal experience, but for me keeping busy and having goals are essential to my well-being. I remember announcing to a friend one day that I was going to see every species of penguin in the world in their natural environment. I don't quite know where this came from, but once I had said it out loud it became a reality, and Mission Penguin was born. I am sure that John Ruskin, who once said 'I find penguins at present the only comfort in life … one can't be angry when one looks at a penguin', would have fully endorsed my mission.

So began the most amazing journey encompassing multiple expeditions around the southern hemisphere. The planning kept me occupied, focused and excited while I saved up the necessary funds for each trip. Where possible I combined destinations, which was financially prudent, accelerated the mission and was better for my carbon footprint.

During my travels I met some incredible people and made new friends, some of whom I subsequently travelled with. I shared my story and was so uplifted by people's support and encouragement. On several occasions additional effort was made to optimise my penguin sightings, for which I am hugely grateful. It was also the response of others that inspired me to share my story more widely, hence this book.

I hope that through my writing and photographs you will join me on my healing journey to some of the most remote places in the world. I also hope that you find inspiration in my story – one of devastation and endings leading to great adventures and new beginnings.

We only have one life, so live it well. Fill it with friends, joy, laughter – and, of course, penguins.

Ursula

OPPOSITE Emperor Penguins, Snow Hill Island, Antarctica.

BELOW King Penguins on the beach at Volunteer Point, East Falkland.

INTRODUCTION

The question that I am most often asked is 'why penguins?'. This is also a question that I ask myself. On a superficial level there is just something about penguins that I find adorable, and judging from the amount of penguin-themed merchandise available I am certainly not alone in this. Penguins capture our hearts, and with their tuxedo-like plumage, amusing waddle and hilarious antics they are the most engaging creatures I have ever come across. In fact, I defy anyone to spend time in the company of penguins and not smile.

For me, however, a love of penguins goes deeper and is more personal. Some of my earliest childhood memories are of my pre-school nursery, which was attached to the local convent. I loved my time there and have such fond memories of the sisters who looked after me so well. They wore the traditional black and white clothing of nuns, and my father, who was not religious, used to refer to them as 'the penguins'. I suspect that deep in my subconscious, my love of penguins began at that very young age.

My next recollection was at age 11, at my junior school. The World Wildlife Fund was running an essay writing competition for schools. My essay was about penguins, and I won the national competition for my age group. I was obviously proud of my achievement and very grateful to my new friends the penguins.

Over time I acquired all sorts of penguin paraphernalia, especially gifts from friends. In fact one friend, Lynn, bought me 50 penguin-themed gifts for my 50th birthday. Luckily, you can never have too many penguins. My husband had also secretly purchased and carefully hidden a silver three-piece penguin cruet set ready for our 25th wedding anniversary. When he had his heart attack, his overriding concern was to share where he had hidden my present in case he didn't make it. Sadly, his fear became reality and I was left all alone to open this most exquisite silver anniversary gift, something that I will treasure for the rest of my life.

During the early 1990s I acquired the nickname 'Pingu' after the fun-loving, mischievous penguin television character created by Otmar Gutmann. This was mainly down to the way that I express excitement by clapping my hands together with straight arms, akin to flippers. This has occasionally caused me problems; for example, when out photographing wildlife and seeing something exciting, which rapidly disappears as my flippers flap. Penguins are, therefore, the perfect subject for me to photograph as not only are they unable to fly off, but they even waddle over to me and join in. Perhaps in penguins I have found my kindred spirit.

OPPOSITE Adélie Penguin, Antarctica.

Meet the
BRUSH-TAILED PENGUINS

Pygoscelis

Although the three species of brush-tailed penguins all look very different, they are linked by each having long, stiff tail feathers, like a brush. They are also three of the four main penguin species found in Antarctica. The Adélie Penguin and Chinstrap Penguin are both ice-dependent, whereas the Gentoo Penguin extends to the subantarctic regions.

All three brush-tailed species form large breeding colonies on ice-free rocky coasts. The nest is a shallow scrape lined with pebbles or other debris to create a small mound for drainage. This helps to protect the eggs and young chicks from meltwater. Competition for pebbles is high, so stealing from other nests is common and can result in high aggression if caught. The females generally lay two eggs, and if there is adequate food then both chicks usually survive. After a few weeks the chicks join crèches, enabling both parents to go out foraging.

Chinstrap Penguin

Adélie Penguin

Gentoo Penguin

OPPOSITE A Gentoo Penguin showing its brush-like tail from which the group are named, Falkland Islands.

CHINSTRAP PENGUIN

CHINSTRAP PENGUIN

Pygoscelis antarcticus

SIZE AND WEIGHT	POPULATION	CONSERVATION STATUS
68–76cm (26.8–29.9in), 3.2–5.3kg (7.1–11.7lb) depending on time of year and gender.	Around 8 million mature individuals and slowly decreasing.	Least Concern (IUCN Red List 2020).

Instantly recognisable by the narrow band of black feathers which pass from ear to ear under the chin, the Chinstrap Penguin has a bluish-black back, crown and tail with a white face and front. It has amber-coloured eyes and pink feet with black soles. The juveniles are similar to the adults but with dark eyes; the chicks are pale grey.

Like many people, as a child I had always associated penguins with snow. In fact, of the 18 penguin species in the world, only four main species are found in Antarctica, and the Chinstrap Penguin is one of them. I was therefore confident that I would see this species on my inaugural trip to kick off Mission Penguin, to the Antarctic Peninsula with Quark Expeditions. I was not disappointed, and in fact the Chinstrap Penguin was the very first new penguin species

I saw. I was just getting over the euphoria of stepping onto the continent of Antarctica for the first time in my life when I spotted a lone individual out on the rocks. I was ecstatic, and my own Pingu-style flippers started flapping in excitement. The penguin stopped, turned and waddled right over to where I was standing. It seemed to be smiling at me, and I couldn't help but reciprocate with the most enormous grin, followed by a few tears.

RIGHT A Chinstrap Penguin clearly showing the origin of its name.

The Chinstrap's name is derived from the narrow band of black face feathers running from ear to ear, which resembles the strap of a black helmet. If I had been naming the species, however, I would have called it the 'Smiling Penguin'. I think that my healing journey really started right there in that moment with that single Chinstrap Penguin.

I am often asked which penguin species is my favourite and, while I love them all, if pressed I would have to choose the Chinstrap Penguin. I am not sure if this is because it was the first species I saw on Mission Penguin, or because of its 'smile', or the way it made me feel on that day. Probably a combination of all three, but this species certainly won my heart.

OPPOSITE My first Chinstrap Penguin. Note the long, stiff tail feathers characteristic of the 'brush-tailed' family of penguins.

PREVIOUS PAGES A lone Chinstrap Penguin at Portal Point, the Antarctic Peninsula.

BREEDING

Chinstrap Penguins are extremely social and breed in very large, densely packed, noisy colonies, usually on rocky slopes. They form long-lasting pairs and return to their natal colony and nest site to breed. The males usually arrive first, in October, closely followed by the females. The nest is made of small stones, which are added to during incubation, and is sometimes lined with feathers and bones. Two eggs are laid a few days apart and incubation is shared by both parents in shifts. If both eggs hatch the two chicks are fed equally. The chicks are brooded for 3–4 weeks, then small crèches form, allowing both parents to forage. The chicks fledge at around two months old.

We also visited Baily Head on Deception Island, which is one of the South Shetland Islands and home to a large breeding colony of Chinstrap Penguins. The colony itself was full of rather scruffy-looking penguins as we visited during the annual moult. Even this, however, could not detract from my joy at the crowd of smiling faces that greeted me. Mission Penguin was up and running and I was absolutely elated.

LEFT Chinstrap Penguins have beautiful, amber-coloured eyes, as well as a gorgeous apparent smile.

BELOW Deception Island is the caldera of an active volcano. Dead krill littered the beach, as they had been literally cooked alive in the warm water of the bay. Krill features in the diet of many penguins. Chinstrap Penguins, along with Adélie Penguins, are actually krill-dependent.

ABOVE A Chinstrap Penguin selects a pebble from the beach, which it will use for nesting. Building nests out of pebbles allows water to drain properly.

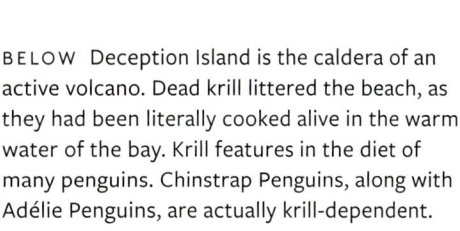

PREY AND PREDATORS

Chinstrap Penguins eat mainly krill, but also some squid and small fish. The eggs and chicks are predated by South Polar Skuas, Southern Giant Petrels and gulls, while the adult birds are targeted mainly by Leopard Seals.

LEFT A Chinstrap Penguin wading across a muddy, guano filled stream.

FAR LEFT The route to the colony includes streams of guano and moulted feathers, so the penguins arrive looking somewhat grubbier than earlier.

OPPOSITE A pristine Chinstrap Penguin arrives on the volcanic beach before beginning its muddy journey to the colony.

RIGHT A Chinstrap Penguin mid-moult.

BELOW Moulting 'smiling' Chinstrap Penguins at Baily Head, Deception Island.

KEY THREATS

Climate change is the greatest threat to the Chinstrap Penguin. Global warming affects its food supply, and the increase in rainfall in Antarctica is proving fatal to the chicks, who are not yet waterproof and can die from hypothermia. Commercial krill fishing and volcanic activity are also significant threats.

LEFT A Chinstrap Penguin having an afternoon snooze.

OPPOSITE The large colony of Chinstrap Penguins at Baily Head, Deception Island.

ADÉLIE PENGUIN

ADÉLIE PENGUIN
Pygoscelis adeliae

SIZE AND WEIGHT
70–73cm (27.6–28.7in), 3.8–8.2kg (8.4–18.1lb) depending on time of year and gender.

POPULATION
Around 10 million mature individuals and thought to be slowly increasing.

CONSERVATION STATUS
Least Concern (IUCN Red List 2020).

Easily recognisable by the white eye-ring, the Adélie Penguin has a white front with a black back, tail and head. Its feet are pink with black soles. Juveniles look similar to the adults but lack the white eye-ring; the chicks are grey.

ABOVE Two Adélie Penguins justifying their reputation as the feistiest of all the penguin species.

OPPOSITE Adélie Penguins use their feet to propel themselves when tobogganing and to steer while swimming. Their flippers help with balance on land.

PREVIOUS PAGES Adélie Penguins in a blizzard, Hope Bay, the Antarctic Peninsula.

The Adélie Penguin is perhaps the most feisty and argumentative of all the penguin species. It was named after Adélie Land, which in turn was named by the French explorer Jules Dumont d'Urville in 1840, in honour of his wife Adéle.

The Adélie Penguin, along with the Emperor Penguin, is the most southerly distributed of all the penguin species and is only found along the coast of the Antarctic continent. Like most expeditions to the Antarctic Peninsula, mine was due to sail down the north-western side before returning via South Georgia and the Falkland Islands. The furthest south we visited was the Yalour Islands, and it was here that I had my first sighting of an Adélie Penguin. It rushed over the snow, seemingly as excited to see me as I was it, but then stumbled and tobogganed for extra speed before eagerly arriving at my feet. I also saw a few other Adélie Penguins there, and although they were not the best sightings I was thrilled to have at least seen and photographed them.

A couple of days later our ship, MV *Sea Adventurer*, suffered mechanical failure in one of its engines. While we could still safely cruise in the vicinity and continued to see wonderful things, it took four days to repair. This had a major impact on our planned itinerary as it meant there was no longer time to visit South Georgia. I was devastated, as South Georgia is known as one of the greatest wildlife spectacles in the world and has three million penguins. However, every cloud has a silver lining, and for us this meant an unscheduled visit to the north-eastern side of the Antarctic Peninsula. We went through the Antarctic Sound overnight and awoke to the most beautiful sunrise and the sea unusually like a millpond. We entered the Erebus and Terror Gulf and marvelled at the ice formations in the early morning sun.

In the far distance on top of a small iceberg I spotted two Adélie Penguins. They had found a superb vantage point from which to enjoy the stunning scenery, and they continued to pose beautifully as the penguin paparazzi arrived in our Zodiac boats to take a closer look.

OPPOSITE Adélie Penguins have stunning white eye-rings and their black feather-covered beaks reduce heat loss.

BELOW Two distant Adélie Penguins on top of a small iceberg.

ABOVE Adélie Penguins moulting at Devil Island.

LEFT An Adélie Penguin with an itch.

OPPOSITE This moulting Adélie Penguin reminded me of a hedgehog. While I was well-wrapped up in hat and gloves to stay warm, the penguins were eating snow to keep cool.

My joy increased even further when we landed on Devil Island and came across several groups of Adélie Penguins in their final stages of moulting. It was here that I took some of my favourite photographs.

All birds moult their feathers and usually this is done just a few at a time. Penguins, however, must remain waterproof, and so once a year they moult all their feathers at the same time, a process known as a 'catastrophic moult'. During this time, which usually takes 2–4 weeks, the birds must remain on land and cannot return to the ocean to feed until the new feathers have pushed through and they are waterproof once again. I watched some of the birds vigorously shaking to accelerate the loss of their moulting feathers and decided that I wanted to capture the spectacle in an image. This took me much longer than I had envisaged, as first I had to find a bird with plenty of space around it and a 'clean background' so that the floating feathers would show up better in the photograph. After finding my subject, I sat down on a rather uncomfortable rock to ensure the right angle for the shot. I then waited and waited and waited. Eventually, after about 45 minutes, the bird finally indulged me and shook, sending feathers flying. Proof, if needed, that patience does pay, and I was thrilled with the result. It was also only just in time, as the transfers back to the ship were well underway by the time I returned, and as usual I was one of the last to leave.

RIGHT An Adélie Penguin shaking to accelerate the moulting process.

As the ship continued, we were told to keep an eye out for stray Emperor Penguins as occasionally they appear on ice floes on this side of the peninsula. When the possible distant sighting came I was so excited, but as we approached it turned out to be a Crabeater Seal (which incidentally do not eat crabs). Had it been an Emperor Penguin I could technically have 'ticked' it off the list, and for a while I was disappointed. Emperor Penguins are, however, the ultimate penguin species and definitely deserve more than a fleeting glimpse of one adult. I also really wanted to see their chicks.

Another expedition would therefore be required to travel even further south into Emperor Penguin territory.

Years later, on my final Mission Penguin expedition in search of Emperor Penguins, I was thrilled to see the Adélie Penguins again on Devil Island. This time it was November, so the beginning of the breeding season, enabling me to observe a large breeding colony. As Adélie Penguins return to the same location to breed, it was wonderful to think that some of the birds I had photographed moulting on my first visit may now be posing for me again.

OPPOSITE The Adélie Penguin colony on Devil Island, Weddell Sea.

BELOW Nest pebbles are a key commodity and squabbles often break out defending them.

BREEDING

Adélie Penguins are monogamous and vigorously defend their nest site in the large, dense colony. The males return to the colony in September or October, closely followed by the females. The nest is a shallow scrape lined with and surrounded by pebbles. The pebbles are important for drainage and are often stolen from other nests, leading to frequent squabbles and even fights. Two eggs are laid a few days apart, with the first egg being the larger of the two. Incubation is shared and lasts for about five weeks. Providing food is plentiful, both chicks will be raised, and crèches form after 3–4 weeks, allowing both parents to forage and provision the chicks. If food is scarce, then only the stronger chick will survive to fledge, usually after around 7–8 weeks.

From the ship, I was also able to photograph penguins both swimming in the Weddell Sea and crossing the sea ice as they went to and from the colony on their foraging trips. I had not seen either of these before as the penguins had been moulting and so not waterproof. I felt so blessed to have had this additional opportunity to see and photograph these delightful feisty penguins.

PREY AND PREDATORS

Adélie Penguins predominantly eat krill, fish and some squid. The eggs and chicks are predated by South Polar Skuas and Southern Giant Petrels, while the adult birds are targeted mainly by Leopard Seals and occasionally Orcas.

OPPOSITE TOP Adélie Penguins 'porpoising' through the Weddell Sea. This enables them to keep breathing without losing momentum.

OPPOSITE BOTTOM Adélie Penguins crossing the sea ice of the Weddell Sea.

BELOW Adélie Penguins have dense plumage and feathered beaks to help keep them warm in the bitter cold of Antarctica.

KEY THREATS

Climate change will likely be the greatest ongoing threat to Adélie Penguins. This affects both their food supply and the extent of the sea ice which is essential to the survival of this species. In addition, there is increasing rainfall in Antarctica, which can have devastating consequences for the young chicks, whose feathers are not yet waterproof. Getting wet and cold can lead to fatal hypothermia.

GENTOO PENGUIN	SIZE AND WEIGHT	POPULATION	CONSERVATION STATUS
Pygoscelis papua	75–90cm (29.5–35.4in), 4.5–8.5kg (9.9–18.7lb) depending on the subspecies, time of year and gender.	Around 774,000 mature individuals and stable overall.	Least Concern (IUCN Red List 2020).

With white freckling and white patches above the eyes on a black head, face and throat, the Gentoo Penguin also has a black back and tail with a white front. Its feet vary from deep pink to yellowy-orange and its beak is bright orangey-red. The juveniles are slightly duller, while the chicks have white fronts, grey backs, and a dull orange beak and feet.

When I started Mission Penguin, the Gentoo Penguin was classed as a single species with a northern and southern subspecies. It has an extremely large breeding range covering both the South Atlantic and Indian oceans. The vast majority, however, breed on the Falkland Islands, Antarctic Peninsula and South Georgia. My Antarctica trip was due to visit all three locations, which meant I would also be able to see both subspecies.

ABOVE My first Gentoo Penguin – a southern subspecies at Yalour Islands (now known as the Antarctic subspecies).

PREVIOUS PAGES Gentoo Penguins returning to feed their chicks, Sea Lion Island, the Falklands.

As we headed first down the north-western side of the Antarctic Peninsula, I had my first sighting of a southern Gentoo Penguin at the Yalour Islands. It was also here that I saw my first Gentoo Penguin chicks, and my heart leapt with delight. The chicks were just adorable, with fluffy grey and white down, orange feet and beaks, and hilarious antics. I could have stayed watching them all day.

On Wiencke Island we were able to disembark the Zodiacs and walk over from Damoy Point to see a small colony of Gentoo Penguins. Photography is so much easier on solid ground – well, relatively solid, as I did stumble and slide in the snow just like the penguins. The chicks in the colony were now in crèches, with just a few adults around to keep an eye on them while their parents were out foraging.

ABOVE A single adult Gentoo Penguin keeping watch over a small crèche.

OPPOSITE A pair of southern (Antarctic) Gentoo Penguin chicks.

PREY AND PREDATORS

Gentoo Penguins eat krill, fish and squid, with the proportions depending on the location and season. The eggs and chicks are predated by Southern Giant Petrels, skuas, Snowy Sheathbills and Kelp Gulls, as well as by Striated Caracaras and feral cats on the Falkland Islands. The adult birds are targeted by Leopard Seals, sea lions, fur seals and Orcas.

The Gentoo Penguin is the third largest of all the penguin species, although the subspecies found in Antarctica is slightly smaller than the one on the Falkland Islands. It also has different face markings with fewer white freckles. I was able to observe this first-hand when I spent an extra week on the Falkland Islands at the end of the expedition cruise and came away with some of my favourite photographs of any penguin species.

There are fewer than 4,000 people living on the Falkland Islands and over one million penguins (of five species), so the stunning beaches were crowded with penguins rather than humans. It was paradise; in fact so much so that I have made two further visits to the islands and taken even more photographs.

ABOVE The beautiful face markings of the northern (Falkland Islands) Gentoo Penguin.

OPPOSITE TOP Gentoo Penguins on the Rookery beach, Saunders Island, the Falklands.

OPPOSITE BOTTOM A Gentoo Penguin nestled in the flowering sea cabbage at Swan Pond, East Falkland.

It can be difficult to resist anthropomorphising penguins. I took photographs of 'laughing' penguins and 'dancing' penguins, to name but a few. I spent hours sitting on beautiful beaches just giggling as I watched their interactions, especially between the adults and chicks. The parents engage their offspring in long food chases before feeding them, to help build up their strength. These are hilarious to the spectator, as the birds charge through the colony as fast as their little feet will carry them, squawking loudly and flippers flapping. The chicks frequently fall over and sometimes even tread on the tail of the adult, making it stumble, before all getting back up and resuming the chase. These pursuits happen simultaneously all over the colony, so it is mayhem, and hard to keep focused on any one family to photograph it – although that may have had more to do with the giggling.

ABOVE Two Gentoo Penguin chicks racing after a parent to be fed.

OPPOSITE These two Gentoo Penguins appeared to be 'laughing' – it was impossible not to join in!

On the Falkland Islands, if Gentoo Penguin chicks are lost the pair can breed again in the same season. This enabled me to see all stages of the breeding cycle in just a few days. I saw Gentoo Penguins mating, collecting pebbles for their nests, with eggs, and even one with a newly hatched chick, with the second egg just cracking open. Skuas were circling constantly and dive-bombing the sitting birds to get them off the nest. Some of these attacks were successful and the penguin eggs or chicks were taken. It was heartbreaking to watch, and I did shed a few tears, but I know as well as anyone that death is part of life (and the skuas had hungry chicks to feed too).

BREEDING

The Gentoo Penguin breeds over an extremely wide geographic range (46°S to 65°S) and the specifics of breeding are therefore highly variable with location. In the more northern populations, the eggs are laid between June and November, whereas further south this is usually between November and December. The circular nest mounds also vary with the available material, so further north may include vegetation and further south predominantly stones, mud and feathers. Two eggs are usually laid a few days apart, and are incubated by both parents for about five weeks. The chicks hatch a few days apart, and both are usually raised and have a good chance of survival if the food supply is plentiful. Crèches form after 3–5 weeks, allowing both parents to forage and provision the chicks. Fledging is at 9–12 weeks in the south but 12–17 weeks in the more northern populations.

OPPOSITE A Gentoo Penguin parent carefully tending to its new offspring.

BELOW LEFT A newly hatched Gentoo Penguin chick resting against its sibling's egg.

BELOW RIGHT Newly hatched, this Gentoo Penguin chick demands food.

ABOVE After a skua successfully dislodged the parent, this Gentoo Penguin chick and hatching egg became highly vulnerable to predation.

LEFT As the sun begins to set, this Gentoo Penguin chick hungrily awaits the return of its parents from foraging.

FAR LEFT A Gentoo Penguin chick looking rather full after its latest feed.

ABOVE Skuas constantly patrol the colony ready to seize any opportunity for food.

It was particularly entertaining watching the adults leaping through the ocean waves as they returned from foraging, and even more amusing trying to guess where they were going to pop up next to photograph them. Gentoo Penguins are thought to be the fastest swimming penguins and can reach speeds of 36km/h (22mph). They seem to 'fly' through the water, but they still need to come up to the surface to breathe. They therefore 'porpoise', leaping in and out of the water like dolphins or porpoises, so can continue breathing without interrupting their forward momentum. What a spectacle, and I was thrilled when my persistence to capture this in a photograph finally paid off.

LEFT A Gentoo Penguin 'porpoising' back to shore.

BELOW A Gentoo Penguin swimming underneath a wave.

OPPOSITE A Gentoo Penguin leaping out from the surf onto the beach.

OPPOSITE A Gentoo Penguin surfing back to shore.

BELOW The Gentoo Penguin is one of the three species known as brush-tailed penguins due to their long, stiff tail feathers. The Falkland Islands beaches were covered in Gentoo Penguin footprints, and it was easy to tell them apart from the other species there due to their dragging feathers leaving a tell-tale trail in the sand – or should that be tell-tail?

RIGHT A lone Gentoo Penguin begins the long walk from the ocean back to the colony.

Staying overnight on the islands meant that I could rise early and return late so immerse myself fully in penguin life from dawn until dusk. What an enormous privilege and joy. My healing journey was well and truly underway and I could not stop smiling.

ABOVE At dawn, a 'waddle' of Gentoo Penguins heads back out to sea to forage for their hungry chicks back at the colony on Saunders Island, the Falklands.

BELOW Gentoo Penguins returning to the colony as the sun begins to set over Sea Lion Island, the Falklands.

ABOVE The setting sun lit up the long, stiff tail feathers of this Gentoo Penguin like a lamp.

BELOW A Gentoo Penguin colony at sunset on Saunders Island, the Falklands.

At the end of 2020, a new scientific study was published suggesting that the Gentoo Penguin should not be regarded as a single species with two subspecies, but rather as four separate species. I heard this reported on the early morning news and was then inundated with emails, texts and calls from my friends to ask what this meant for Mission Penguin. I searched online to find the full research paper and frantically read through pages of analysis to discover more. The proposal was to have an Antarctic species, a Falkland Islands species, a South Georgia species and a Kerguelen Island species. This is not just of interest academically, but also has important consequences for conservation. The global population of Gentoo Penguins is thought to be stable and is classified as Least Concern on the IUCN (International Union for Conservation of Nature and Natural Resources) Red List (2020). Within this, however, some populations are increasing while others are decreasing at such a rate that, if classified as a separate species, they may meet the Vulnerable criteria and warrant specific conservation measures.

I knew that I had seen and photographed the Antarctic and Falkland Islands Gentoo Penguins and by an amazing piece of luck I had also seen and photographed a few

KEY THREATS

While the global numbers of Gentoo Penguins are thought to be stable, there is wide variation in the different populations, with those on the Antarctic Peninsula and the Falkland Islands increasing while those on South Georgia and the Kerguelen Islands are decreasing. Reasons for these declines include toxic algal blooms, disease, competition for food (both natural, for example from increasing fur seal populations, and from commercial fisheries), oil and plastic pollution and of course climate change, which affects food location and abundance.

BELOW A lone South Georgia Gentoo Penguin with a group of King Penguins on the beach at Stromness, South Georgia, with the remains of the whaling station behind.

Gentoo Penguins on South Georgia. This was en route to Tristan da Cunha, where Mission Penguin took me in search of Moseley's Rockhopper Penguin.

I had not, however, visited the Kerguelen Islands in the Indian Ocean, home to the proposed fourth new species. The closest I had got was Macquarie Island, where I had seen and photographed a few Gentoo Penguins while searching for the Royal Penguin.

I reread the paper to see where the Macquarie Gentoo Penguin fitted, and discovered that it was linked to the Kerguelen Island species. On contacting the lead author, this was confirmed, much to my relief! Mission Penguin was future-proofed, at least for now, although I am not entirely ruling out another expedition sometime to some of these Indian Ocean islands.

In 2022 the proposed four species were officially listed as Gentoo Penguin subspecies in the established authority *Clements Checklist of Birds of the World*.

TOP RIGHT A Macquarie Island Gentoo Penguin family (closely related to the Kerguelen Island subspecies).

RIGHT Two Macquarie Island Gentoo Penguin chicks begging to be fed. The chicks are similar size suggesting plentiful food supply.

BELOW Gentoo Penguin colony at Seal Bay, East Falkland.

Meet the
BANDED PENGUINS
Spheniscus

African Penguin

Humboldt Penguin

Galápagos Penguin

Magellanic Penguin

The banded or ringed penguins are the four most northerly penguin species and are found in South America and southern Africa. All four look very similar, with black and white bands running down their face, neck and chest. The only other colouring is a patch of bare pink skin between the eye and the beak. This area is highly vascularised to help them lose heat when required in these warmer climates. Their more temperate habitats mean that the banded penguins are the species most commonly found in zoos around the world, as well as being the most accessible to tourists. Their closer proximity to civilisation also puts them under pressure from human activity.

All four species are monogamous and nest in cavities or burrows, which also helps them to escape the heat of the day. When food supplies are plentiful they lay two eggs, with both chicks likely to survive, whereas if food is short they lay just one egg.

OPPOSITE The banded family of penguins are sometimes known as the 'Jackass penguins' due to their distinctive 'braying' call which sounds like a donkey. Here a pair of Magellanic Penguins bray together on Saunders Island, the Falklands.

AFRICAN PENGUIN

AFRICAN PENGUIN *Spheniscus demersus*	SIZE AND WEIGHT 60–70cm (23.6–27.6in), 2.1–3.7kg (4.6–8.2lb) depending on time of year and gender.	POPULATION Around 41,700 mature individuals and rapidly decreasing.	CONSERVATION STATUS Endangered (IUCN Red List 2020).

With a single black band across the upper part of the chest, the African Penguin has a black back and tail, a white front with black flecks, and black feet with occasional pink patches. It has a black and white face with an area of bare pink skin between the eyes and black beak. The juveniles have blue-grey backs and lack the chest band, and the chicks are brown and white.

Just as the African Penguin was the first penguin species to be observed by modern Europeans in 1497, it was also the first species that I saw in the wild, albeit about 500 years later.

I had been to South Africa before my husband died, so this was one of three penguin species that I had seen before Mission Penguin began. My first visit there was in the days of film rather than digital cameras and I took some lovely photographs down on the sand at Boulders Beach near Simon's Town. There were very few people around and you could walk to the beach, sit down, and before long the penguins would just wander over to you to investigate. It was a magical experience and one that cemented my love for penguins, which until then had been based solely on books and television documentaries. Having seen this penguin species already I did not actually need to make another trip to South Africa; however, when some friends suggested going to Cape Town to celebrate a birthday I jumped at the chance. My only stipulation was that we would also go and visit the penguins while we were there.

LEFT The African Penguins were initially described by the Portuguese navigator Vasco da Gama as 'Jackass Penguins'. Long before they were seen and identified as birds, they were heard braying like donkeys on the southern African shore.

OPPOSITE An African Penguin clearly showing the black and white banding that makes it one of the four species known as the banded or ringed penguins.

PREVIOUS PAGES A pair of African Penguins at Stony Point, Betty's Bay, South Africa.

What I hadn't realised when I first went to Boulders Beach was that it was a relatively new colony. It was formed in the early 1980s when a pair of African Penguins found a safe area to raise their chicks between the sea and the houses just beyond the beach. The colony was still relatively small when I first visited, but there are now over 2,000 birds. This has also led to an explosion in visitor numbers, which now exceed 60,000 people each year. To minimise disturbance and protect the penguins there are, therefore, new walkways and ropes to restrict access to certain areas, and the whole experience was much more commercialised. Despite this, I could still get close enough to the penguins to fully enjoy their antics and see their beautiful markings.

We also visited a second colony at Stony Point, Betty's Bay. This was less busy than Boulders Beach and enabled me to spend more time enjoying these fabulous penguins as they returned from the sea, most likely from foraging trips.

ABOVE An African Penguin coming back to the colony from the ocean.

OPPOSITE Other than black and white, the only colouring on the African Penguin is the patch of bare pink skin between the eye and the beak. The area is highly vascularised to help prevent overheating in the hot African sun.

PREY AND PREDATORS

African Penguins predominantly eat shoaling fish such as anchovies and sardines, with some squid and small crustaceans. At sea they are predated by Cape Fur Seals and sharks, while on land their eggs and chicks are killed by dogs, feral cats, Kelp Gulls and rats.

The African Penguin is unusual in that it can breed throughout the year, so I was able to see eggs, chicks, juveniles and adult birds during my visit. I even managed to see mating penguins, which is quite a spectacle. Penguins have evolved torpedo-shaped bodies that are perfect for 'flying' through the water but are not quite so suited to reproduction. It is like balancing a barrel on top of another one. The male frequently rolls off so must keep climbing back on and trying again. Luckily, they weren't bashful, and it provided hours of entertainment. African Penguins also mate for life and there was lots of bonding behaviour on show, which appealed to my romantic nature and really lifted my spirit. Love was certainly in the air.

BREEDING

The African Penguin can breed at any time of year and the preferred nest site is a burrow excavated in guano or firm sand. Two eggs are usually laid up to three days apart, and are incubated by both parents for about six weeks. Providing there is sufficient food, both chicks are raised. They are brooded by both parents for around four weeks, then form small crèches, enabling both parents to provision the hungry chicks. Fledging varies between two and four months, and the immature birds develop adult plumage over two years.

BELOW Penguins swim underwater using their flippers for propulsion and their large feet as rudders for steering.

OPPOSITE Balancing during mating can be challenging. Persistence finally pays off and their two cloaca (reproductive and waste orifice) align, enabling sperm to be transferred.

When I first visited Boulders Beach it was free, but there is now a fee at both this colony and at Betty's Bay. The funds raised go towards conservation, which is essential as this species seems to be rapidly declining and may soon move from Endangered to Critically Endangered on the IUCN Red List.

TOP LEFT A shortage of vegetated habitat means that some penguins nest in the open, which can lead to fatal hyperthermia. This may increase with global warming.

LEFT Two African Penguins relaxing together.

BELOW To avoid the heat of the sun, African Penguins nest in shallow scrapes under vegetation or in burrows.

KEY THREATS

The African Penguin population is undergoing a rapid decline, most likely due to commercial fishing and climate change. Warming sea temperatures shift the prey populations so the distance between the feeding grounds and colony are too great, leading to mass starvation and breeding failure. The proximity of many colonies to commercial shipping ports also increases the risk of both chronic oil pollution and large individual spills. New ports as well as a demand for coastal housing development lead to habitat loss, and there is ongoing competition for food and breeding space with Cape Fur Seals.

RIGHT A pair of African Penguins with their immature offspring, which lacks the banding of the parents.

HUMBOLDT PENGUIN

HUMBOLDT PENGUIN *Spheniscus humboldti*	SIZE AND WEIGHT	POPULATION	CONSERVATION STATUS
	65–70cm (25.6–27.6in), 4–5kg (8.8–11lb) depending on time of year and gender.	Around 23,800 mature individuals and decreasing.	Vulnerable (IUCN Red List 2020).

The Humboldt Penguin has a white front with black flecks and a single black band across the top. It has a black back and tail and a black and white head with a large area of pink skin. This runs from the eyes to the black beak and continues under the chin. The black feet may also have pink patches. The dark grey and whitish juveniles lack the chest band, and the chicks are greyish-brown and white.

The Humboldt Penguin is named after the cold-water current it swims in, which is itself named after the Prussian naturalist and explorer Alexander von Humboldt. The Humboldt Penguin is found along the coastlines and associated small islands of Peru and Chile. Most of these are not easily accessible and are certainly not on the main tourist route. Having researched a relatively accessible location I contacted a travel company to help me with the logistics. The agent was initially concerned and explained that there were fewer penguins there than there used to be. They estimated my chance of success at only 50 per cent and they didn't want me to be disappointed. After sharing about Mission Penguin and explaining that not going at all reduced my chances to zero per cent, they pulled together a fabulous itinerary.

Travelling with a friend, we landed in Lima, Peru, in the early morning and were taken to a hotel for our first night. Unfortunately our room wasn't ready and, exhausted from travelling, we both fell asleep on a large sofa in the reception area. The staff must have wanted us to feel at home because they searched out some English recorded music for us to listen to while we were waiting. One of my abiding memories was waking up in the middle of April to a loud rendition of the Christmas carol 'Hark! The Herald Angels Sing'. It was so touching that they had gone to such trouble that we didn't have the heart to explain it wasn't quite the right time of year.

The location I had selected to search for the Humboldt Penguin was the Ballestas Islands. This entailed a long car journey from Lima down dusty and bumpy roads to the town of Paracas on the south coast of Peru. From there I would take a boat trip out to the islands the following day. I awoke early the next morning feeling excited but also increasingly nervous. Perhaps the travel agent had been right and I would not get to see the penguins after all. As soon as the islands came into distant view, I raised my binoculars and frantically scanned the mass of seabirds covering them.

Suddenly, as the boat rose on a wave, I caught a glimpse of two Humboldt Penguins among a large colony of Peruvian Boobies. As we got closer, I even managed a couple of photographs from the bobbing boat and, although they were not the best, I could at least then relax and enjoy the rest of the trip – mission Humboldt accomplished. As the boat tour continued, I saw only three other penguins, so just five in total. Not exactly a colony, but enough to make me smile and let out a huge sigh of relief. I hoped the reason there were so few penguins there was that they were all out fishing.

OPPOSITE The Ballestas Islands off the south coast of Peru used to be around 10m (32ft 10in) higher, before guano mining reduced them back to their bedrock.

PREVIOUS PAGES Two Humboldt Penguins among a large colony of Peruvian Boobies on the Ballestas Islands of Peru.

PREY AND PREDATORS

Humboldt Penguins eat a variety of fish including Peruvian Anchovy, Peruvian Silverside and Araucanian Herring, plus some squid. At sea they are predated by Orcas and South American Fur Seals, while on land eggs and chicks are killed by dogs and feral cats, Kelp Gulls, Turkey Vultures and rats.

ABOVE Two Humboldt Penguins preening on the rocky shoreline, perhaps after just returning from a foraging trip.

LEFT A Humboldt Penguin clearly showing its black and white bands and the exposed highly vascularised pink skin on its face.

BREEDING

The Humboldt Penguin has two main breeding seasons, depending on the location. They usually nest in a burrow excavated in ancient guano. Two similarly sized eggs are usually laid 2–4 days apart, and after six weeks of incubation by both parents also hatch around 2–4 days apart. Both chicks are usually raised and stay in the burrow for the first 2–3 weeks, with the parents alternating between brooding and foraging. The chicks do not form crèches, so once old enough they are left unattended, allowing both parents to forage. They usually fledge between 10 and 13 weeks.

I learned, however, that the main reason there are so few Humboldt Penguins today is the historical practice of guano mining. Guano is the excrement of seabirds which accumulates over centuries and forms a soft substrate for the penguins to burrow into for nesting.

Unfortunately for the penguins, guano is also an incredibly rich natural fertiliser, and became a high-value commodity to Peru in the 1800s. Intensive guano mining continued for around 40 years, removing millions of tonnes of this precious material.

The excavation of this vital nesting substrate, leaving only impenetrable bare rock, decimated the Humboldt Penguin population by over 90 per cent. Interestingly, it was a London-based company, Antony Gibbs & Sons, who acquired the rights to supply Peruvian guano to the North American and European markets, with Great Britain a major customer. William Gibbs, one of the sons, became extremely wealthy as a result and bought then remodelled Tyntesfield, which is now a National Trust property near to where I live in the south-west of England.

KEY THREATS

The Humboldt Penguin population is highly susceptible to El Niño events, which are predicted to increase in intensity and frequency with climate change. During these events, nests can flood and the sharp rise in sea temperature reduces prey availability, leading to mass starvation and breeding failure. Commercial fishing, especially gillnet fishing, is an ongoing threat, as is the potential for oil pollution.

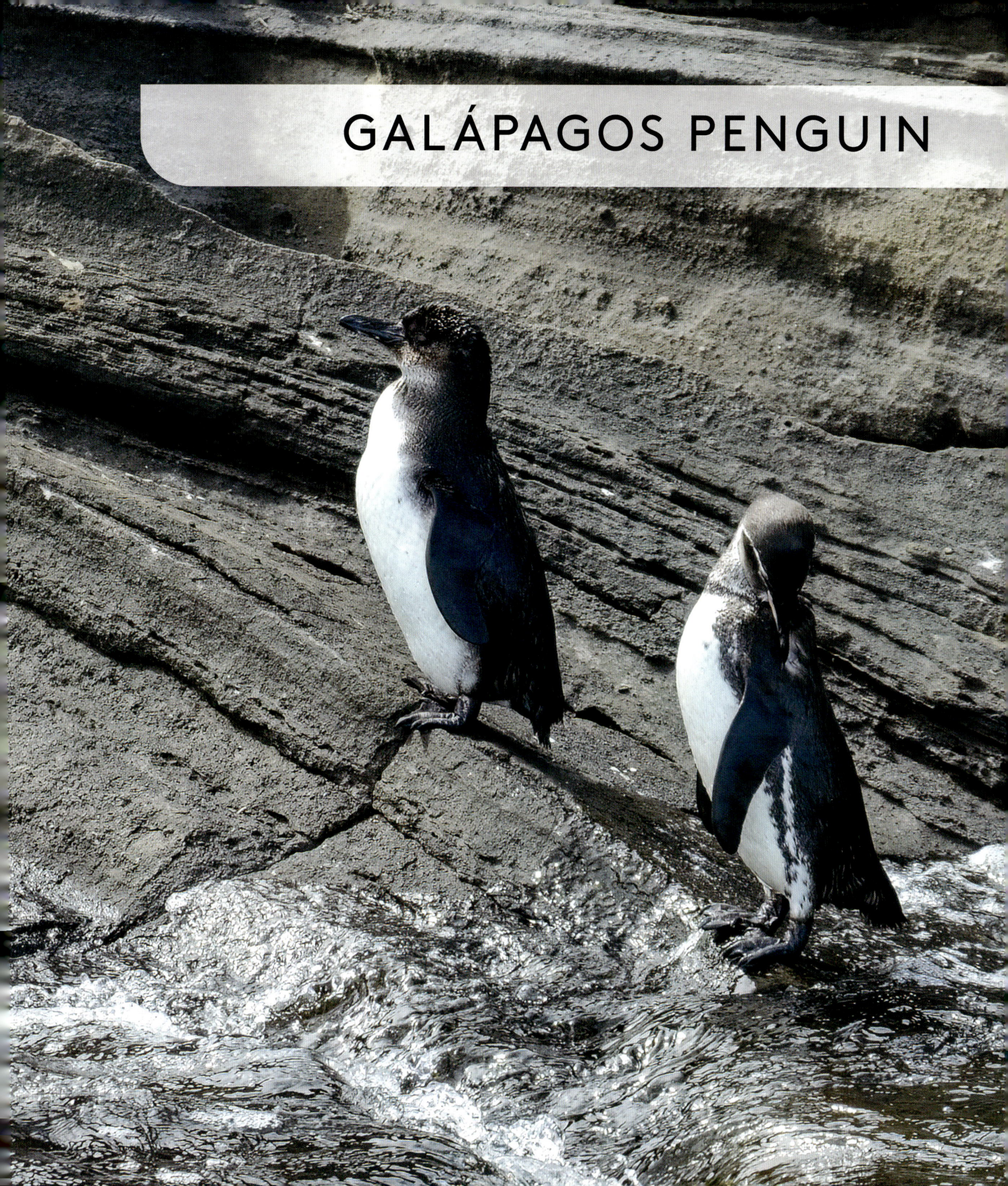

GALÁPAGOS PENGUIN

GALÁPAGOS PENGUIN *Spheniscus mendiculus*	SIZE AND WEIGHT 48–53cm (18.9–20.9in), 1.4–2.9kg (3.1–6.4lb) depending on time of year and gender.	POPULATION Around 1,200 mature individuals and decreasing.	CONSERVATION STATUS Endangered (IUCN Red List 2020).

The smallest of the banded penguins, the Galápagos Penguin has the least distinct upper breast-band on a white front flecked with black. The rest of the body is black other than a thin white stripe curving from each eye to the throat and bare pink skin at the base of the beak. The black feet may have white patches. Dark grey and white juveniles lack the breast-band; chicks are greyish-brown and white.

At school my favourite subject was biology, and I loved learning about Charles Darwin's visit to the Galápagos in 1835 on the HMS *Beagle* expedition. Darwin noticed that each volcanic island had its own unique species; for example, of finches, mockingbirds and giant tortoises. From these observations he developed his theory of evolution, published in 1859. At school I never dreamed that one day I would follow in his footsteps, but as the Galápagos have their own endemic species of penguin I had to go there.

Despite there being about 20 islands in the archipelago, the Galápagos Penguins are predominantly found on just two: Isabela and Fernandina, in the cooler waters of the western side. This meant carefully choosing an itinerary, from the many available options, that specifically visited these islands. The selected boat, the MY *Letty*, could accommodate just 20 passengers, which enabled fabulous sightings and numerous opportunities for snorkelling, cruising and walking tours on the various landings. It was a truly spectacular trip, and my camera was busy from dawn until dusk with various species of booby, frigatebird, both Galápagos Land and Marine Iguanas, and of course the giant tortoises, to name just a very few.

My main target, however, was the Galápagos Penguin, and I had to wait five days until we finally landed on Fernandina. Fernandina is the newest of the volcanic islands and predominantly a black lava landscape. Here I saw hundreds of Marine Iguanas, numerous lizards and the Galápagos Flightless Cormorant, all of which were wonderful, but no penguins. I tried to keep calm; after all, the stronghold for the breeding colony was Isabela Island, which we were due to visit the next day.

ABOVE A small colony of Galápagos Penguins breed on this islet just off the coast of Isabela Island.

PREVIOUS PAGES An adult Galápagos Penguin preening with a juvenile and a Marine Iguana, also unique to the islands.

BREEDING

The Galápagos Penguin can breed at various times of year providing the sea temperature is lower than 24°C (75°F), to optimise food supply. They create a simple nest of twigs in natural cavities such as lava crevices or underneath boulders to shelter from the equatorial sun. Two eggs are laid a few days apart and incubated for about six weeks, and after hatching both chicks are raised. The chicks are brooded and fed alternately by both parents for the first 30 days and are then left close to the nest site, allowing both parents to forage. They fledge at around nine weeks.

At the beginning of the cruise, I had shared Mission Penguin with the crew and my fellow passengers, so everyone knew of my desperation to see at least one Galápagos Penguin. To increase my chance of success one of our local guides secured special permission to cruise around a small islet just off Isabela Island. This was not on the tourist route but was a known breeding site, and it was here that I had my first sighting. The penguins were scrambling up the volcanic rock, still wet from the ocean and in full sun which, while lovely to see, was a nightmare to photograph. Most of my shots gave them overexposed white fronts, but even that didn't matter – I had seen the Galápagos Penguin and I was thrilled.

ABOVE The juvenile Galápagos Penguin lacks the breast-band of the adult and has a greyer back and head without the exposed facial pink skin.

LEFT The banding of the Galápagos Penguin is less distinct than the other three species in the group and it is also the smallest of the four banded penguins.

PREY AND PREDATORS

Galápagos Penguins eat a variety of schooling fish including anchovies and sardines. At sea they are predated mainly by sharks, while on land the adults and juveniles are targeted by feral cats and dogs. The eggs and young chicks are also killed by snakes, owls, crabs and rats.

Later that day, while cruising to admire the mangrove vegetation on the shoreline of Isabela Island around Elizabeth Bay, I spotted a few distant penguins out foraging. I even managed to photograph one that was a little closer, although unfortunately it was swimming in the opposite direction.

With the Galápagos Penguin achieved, I could relax and marvel at the rest of the amazing flora and fauna of these spectacular islands. I was particularly relieved as the Galápagos Penguin is the rarest and most endangered of the penguin species, so I felt incredibly lucky to have seen it. It was a truly memorable expedition and a childhood dream come true.

BELOW Penguins are only found in the wild in the southern hemisphere. Breeding close to the equator, the Galápagos Penguin is therefore the most northerly species of penguin. When foraging for food, however, they sometimes swim over the invisible 'line', making them the only penguin to be found naturally in the northern hemisphere (even if only temporarily). A useful fact if you enjoy quizzes!

KEY THREATS

The Galápagos Penguin is highly susceptible to strong El Niño events, which lead to fatal food shortages and accounts for the widely fluctuating population size. After a major El Niño event in 1982/3 it was estimated that just 700 individuals survived. The projected increased frequency and intensity of these events due to climate change has led to predictions that the Galápagos Penguin could be the first penguin species to go extinct in the wild. Living on highly active volcanic islands also brings the threat of whole colonies being wiped out by lava flows reaching the coast. Other threats include local fishing and pollution as both the resident human population and tourism continue to grow.

MAGELLANIC PENGUIN

MAGELLANIC PENGUIN	SIZE AND WEIGHT	POPULATION	CONSERVATION STATUS
Spheniscus magellanicus	Approximately 70cm (27.6in), 2.3–7.8kg (5.1–17.2lb) depending on time of year and gender.	2.2–3.2 million mature individuals and decreasing.	Least Concern (IUCN Red List 2020).

The Magellanic Penguin is the only banded penguin to have two black bands visible from the front. The back and tail are black and the white front is lightly flecked with black. Its black and white face has bare pink skin from the eyes to the top of the black beak. The feet are black with pink patches. Juveniles have grey backs and pale cheeks with no chest banding; chicks are brownish-grey and white.

The Magellanic Penguin is named after the Portuguese explorer Ferdinand Magellan, whose crew spotted it in 1520 as they sailed around the tip of South America during the first circumnavigation of the Earth. Over 90 per cent of the total Magellanic Penguin population breed in coastal Patagonia, with the remainder on the Falkland Islands. I decided to focus my attention for Mission Penguin on the smaller Falkland Islands group as this would also give me access to another four penguin species during the same trip.

My voyage to the Antarctic Peninsula with Quark Expeditions culminated in two days at the Falkland Islands. We went ashore via the Zodiacs at Volunteer Point, a headland on the east coast of East Falkland. This has the most beautiful, although somewhat windy, white sandy beach and it was here that I had my first sighting of Magellanic Penguins. I followed them as they waddled up the beach and over the dunes until they reached an area full of burrows that they use for nesting. Some of the burrows were occupied, but as the

LEFT Magellanic Penguins on the 'runway' of Saunders Island, the Falklands.

OPPOSITE Magellanic Penguins nest in deep burrows, which can also protect them from the heat of the sun and the strong Falkland Islands winds.

PREVIOUS PAGES Magellanic Penguins on Saunders Island, the Falklands, clearly showing the classic penguin 'countershading'. Their black backs help to avoid detection from above by both prey and predators, while their white fronts camouflage them from beneath.

breeding season had finished, I think that they were sensibly using them to shelter from the wind. Sand is the enemy of photographic equipment, and I could have done with a burrow myself at times.

Our ship expedition ended and we disembarked for the last time at Stanley on East Falkland, but I opted to continue Mission Penguin by staying on for a further week to find some of the other penguin species that breed on the Falkland Islands. With well over a million penguins there, covering five species, you can imagine my continued excitement as one expedition ended and a new one began.

As well as the main East and West Falkland islands there are another 776 smaller islands and islets, although very few of these are inhabited let alone have available tourist accommodation. Onward travel from East Falkland is via small red FIGAS (Falkland Islands Government Air Service) planes, which is a wonderful experience in itself. They are basically the Falkland Islands taxi service, used to travel between the islands. Each island is run by an owner or manager and it is their responsibility to keep clear a strip of land that the small planes can use for take-off and landing. One of my most memorable landings was when the pilot had to circle several times at low level to 'buzz' a group of Magellanic Penguins that were enjoying the short grass of the 'runway'. They took a little persuading to waddle off and make room for our plane to land, but luckily were soon back to welcome the new arrivals. The hosts were also there to greet us, but could not compete with the penguins' salutation. The birds even returned to see us off two days later, requiring the pilot to 'buzz' them once again before he could collect us.

It is difficult to convey the spectacular beauty and remoteness of the Falkland Islands. The limited accommodation can sometimes mean you are the only guests staying and have the whole island to yourself – well, except for the penguins of course. I am not exaggerating when I say that the Falkland Islands have become my favourite travel destination in the world.

It was not surprising, therefore, that I seized upon one last opportunity to visit the islands while writing up Mission Penguin. I travelled in mid-January, during the breeding season, so was able to photograph the gorgeous chicks as well as more adult birds.

The chicks were at varying stages of development, with some just starting to emerge from their burrows. These youngest chicks were still in their brown and white fluffy attire, while others were at the halfway point and some had even fully transformed into their juvenile plumage.

BREEDING

Magellanic Penguins nest in deep burrows dug into a variety of substrates, or occasionally in shallow scrapes underneath bushes, tree roots or rocks. The males usually return to the colony in September with the females arriving shortly afterwards.

There is high pair-bond and nest-site fidelity. Two eggs are laid about four days apart, with the first egg being larger than the second. Incubation is shared and lasts for about six weeks. The eggs usually hatch one day apart and the parents alternate brooding and foraging for about four weeks. The chicks are then left unattended in the burrow and fledge at 9–17 weeks, depending on the food supply. In ideal conditions both chicks are raised, but in times of food shortage the first-hatched chick is preferentially fed.

ABOVE Safely hidden in their burrow, these two Magellanic chicks wait for their parents to return from foraging.

LEFT Young Magellanic Penguin chicks have brown and white feathery down which blends in with the earth around the burrows, camouflaging them from predators.

OPPOSITE The juvenile Magellanic Penguin lacks the banding of the adults.

LEFT The call of the Magellanic Penguin sounds like a braying donkey.

OPPOSITE Two Magellanic Penguin parents 'braying' together after one returns from foraging. After a brief reunion, one stays to feed the hungry chicks and the other departs.

The adult birds returned to their burrows after foraging and stood outside 'braying' to announce their arrival to the hungry chicks below ground. There then began the most hilarious food chases, with the chicks desperate to outrun each other by whatever means possible, fair or foul, to be fed first. It was difficult to keep the camera steady, I was laughing so much.

PREY AND PREDATORS

Magellanic Penguins eat predominantly anchovies and sardines, with some crustaceans and cephalopods. The eggs and chicks are predated by skuas, gulls and caracaras, and occasionally foxes, feral cats and dogs, depending on the location. At sea and on land the adults are targeted mainly by South American Sea Lions and giant petrels.

TOP RIGHT Magellanic Penguin chicks compete for food. In times of plentiful supply, both chicks are provisioned equally.

RIGHT Magellanic Penguin chicks cajoling the parent into regurgitating more food.

OPPOSITE A Magellanic Penguin parent leading its two offspring on a food chase. This builds up the chicks' strength before they are fed.

RIGHT As the sun rises on the horizon, two Magellanic Penguins set out to forage.

OPPOSITE Magellanic Penguins are monogamous and remain faithful to each other over many breeding seasons. Nesting underground in burrows means they can get rather muddy.

Magellanic Penguins are found on most islands in the Falklands with accessible beaches, and I have been lucky enough to see them on all seven islands that I have visited so far – only another 771 islands to go.

KEY THREATS

The primary threats facing Magellanic Penguins are oil pollution, food competition with fisheries, bycatch and climate change.

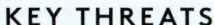

Meet the
CRESTED PENGUINS

Eudyptes

As the group name suggests, all these seven species have head-crests, which vary in the shade of yellow, position on the head and length. The least extravagant of these belongs to the Southern Rockhopper Penguin, and interestingly the most extravagant is in Moseley's (or Northern) Rockhopper Penguin, which used to be a subspecies of the former. The Royal Penguin and Macaroni Penguin have the brightest yellowy-orange headgear compared to the paler yellow of the other species. Six of the species look similar, with a black face and throat, whereas the Royal Penguin has a white face and throat. All the species share red eyes, a sturdy orangey-red beak, and pink feet with dark soles.

The crested penguins all nest on islands north of the Antarctic Convergence Zone. Some have a broad range; for example, the Macaroni Penguin is found in both the South Atlantic and Indian oceans. Other species, however, have an extremely narrow range; for example, the Snares Penguin only breeds on the Snares Islands.

All the species scale steep rugged terrain to access inland colonies for nesting; some of these colonies are among forests, such as those of the Fiordland Penguin. Perhaps the most famous of the crested penguins are the rockhoppers, which, as their name suggests, hop and jump their way up cliff faces.

Most of the crested penguin species are monogamous. They generally raise only one chick, despite laying two eggs. Interestingly, the second egg is usually much larger than the first and, although laid later, if both eggs are incubated it hatches first. It is then the chick from this second egg that the parents focus on, unless it is lost, and it is therefore this chick that usually survives to fledge. Several theories have been put forward to explain this unusual behaviour, but the mystery remains.

BELOW Macaroni Penguins in various stages of moulting at Hercules Bay, South Georgia.

Macaroni Penguin

Southern Rockhopper Penguin

Moseley's (or Northern) Rockhopper Penguin

Fiordland Penguin

Snares Penguin

Royal Penguin

Erect-crested Penguin

MACARONI PENGUIN

MACARONI PENGUIN

Eudyptes chrysolophus

SIZE AND WEIGHT

Approximately 71cm (28in), 3.1–6.6kg (6.8–14.6lb) depending on time of year and gender.

POPULATION

12.6 million mature individuals and rapidly decreasing.

CONSERVATION STATUS

Vulnerable (IUCN Red List 2020).

The Macaroni Penguin's head-crest is bright orange-yellow streaked with black, starting at the forehead and flaring back over the eyes. It has a bluish-black back, tail, head, face and throat and a white front. It has an orange-brown beak with some visible skin at the base and pink feet with black soles. The duller juveniles lack the crest and have a greyish throat and the chicks are brown and white.

Although rapidly declining, the Macaroni Penguin is thought to be the most common penguin species in the world, and it is also one of the most widespread, with huge colonies covering both the South Atlantic and Indian oceans. It should therefore have been one of the easiest species to see and photograph – or so I thought.

While numbers are very low on the Falkland Islands, there are about two million Macaroni Penguins on South Georgia, so I felt confident about seeing them when I booked my month-long expedition to the Antarctic Peninsula, South Georgia and the Falkland Islands.

My first sighting was on Livingston Island, part of the South Shetland archipelago just north of the Antarctic Peninsula. It was only one bird but still technically a 'tick' for Mission Penguin, so I celebrated that evening, with the hope of so many more to come on South Georgia.

RIGHT A lone Macaroni Penguin, Livingston Island, South Shetlands.

It was at this point that our expedition ship broke down and South Georgia was removed from the itinerary. I was very disappointed on two counts. Firstly, and most obviously, I would not get to see the two million Macaroni Penguins. Secondly, I had always been fascinated by the incredible 1916 *Endurance* rescue mission from Elephant Island to South Georgia led by Sir Ernest Shackleton and I was keen to see the terrain for myself. The only consolation was that at least I had seen one Macaroni Penguin on the trip and there was still a remote possibility of seeing some of the few individuals that breed on the Falkland Islands. Although this number is very small and there are 778 Falkland Islands in total, incredibly I struck lucky. While on Pebble Island observing a colony of moulting Southern Rockhopper Penguins, I spotted four conspicuous individuals with much brighter yellowy-orange plumes. They were also moulting so rather scruffy in appearance, but definitely Macaroni Penguins, and I was overjoyed at the sight. I did find it rather ironic, however, that with the Macaroni Penguin thought to be the most common penguin in the world, I had only managed to see five of them.

ABOVE Two pairs of moulting Macaroni Penguins among a whole colony of moulting Southern Rockhopper Penguins on Pebble Island, the Falklands.

PREVIOUS PAGES Macaroni Penguins at Hercules Bay, South Georgia.

Although I still desperately wanted to visit South Georgia, it was difficult to justify an additional trip when I had already seen the four penguin species that breed on the island. I was, therefore, thrilled to discover that my expedition to Tristan da Cunha a few years later in search of Moseley's Rockhopper Penguin also spent several days at South Georgia. It was here, at Hercules Bay, just north of Stromness, that I finally got to see a colony of Macaroni Penguins at their breeding site. Once again, I had missed the actual breeding season and the annual moult was already underway, but it was wonderful to finally see a greater number of them.

BREEDING

Macaroni Penguins breed in very large, dense, noisy colonies which can be over 100m (328ft) above sea level. These are usually on rocky terrain, but can also be among tussock grass if it is available. The nest is a shallow depression lined with mud and small stones. The exact breeding season varies with location, but the two eggs are usually laid in November, 4–5 days apart. The first egg, which is smaller, is usually lost during the five-week incubation, leaving only the second egg to hatch. The chick is brooded by the male for around three weeks while the female forages and provisions it. Small crèches then form, allowing both parents to forage until the chick fledges at 9–10 weeks.

South Georgia was easily my favourite part of that expedition, and my camera was very busy. About 30 million seabirds breed there, including over three million penguins (of four species), as well as several million Antarctic Fur Seals. The Antarctic Fur Seal pups were extremely cute, but while the dramatic recovery of the population following the end of their historical exploitation is a huge success story, it does have serious consequences for the penguins. Fur seals will predate penguins, but more importantly they compete for food and can block access to breeding sites. This may be one of the key reasons behind the huge decline in the numbers of Macaroni Penguins over the last three decades.

ABOVE Antarctic Fur Seal pup, South Georgia.

RIGHT The Macaroni Penguin was named by British explorers who likened them to eighteenth-century young gentlemen famous for their flashy dress, including hats adorned with colourful feathers known as 'Macaronis'.

PREY AND PREDATORS

Macaroni Penguins eat predominantly krill, fish and some squid. The eggs and chicks are predated by skuas, gulls, giant petrels and sheathbills. At sea the adult birds are taken by Leopard Seals and Antarctic Fur Seals.

KEY THREATS

The primary threats facing Macaroni Penguins are commercial fisheries, climate change, disease, competition for space and prey with fur seals, and also natural disasters, such as volcanic eruptions, which can wipe out entire colonies.

My fortuitous visit to South Georgia was just paradise. As well as delighting in the penguins, I also fulfilled my desire to visit Shackleton's grave and I did have a 'wee dram' of his favourite tipple to toast his memory.

Several years later I also managed to revisit the Falkland Islands during the breeding season, so saw a few more adult birds and finally photographed one with a chick.

LEFT A Macaroni Penguin and chick, Kidney Cove, East Falkland.

OPPOSITE A Macaroni Penguin, Kidney Cove, East Falkland, with its bright yellowy-orange and black plumes that start from the centre of the forehead and sweep backwards.

SOUTHERN ROCKHOPPER PENGUIN

SOUTHERN ROCKHOPPER PENGUIN
Eudyptes chrysocome

SIZE AND WEIGHT	POPULATION	CONSERVATION STATUS
45–58cm (17.7–22.8in), 2–3.8kg (4.4–8.4lb) depending on time of year and gender.	Around 2.5 million mature individuals and decreasing.	Vulnerable (IUCN Red List 2020).

The narrow pale yellow crest of the Southern Rockhopper Penguin starts as a thin stripe over each bright red eye and increases in length towards the back of the head. It has a black back, tail, head and throat with a white front. The beak is orange-red and the feet are pink with black soles. The duller juveniles have a minimal crest and grey throat; the chicks are brown and white.

While researching where was best to see the Southern Rockhopper Penguin I watched a documentary about the Falkland Islands that included footage of them queueing up to have a freshwater 'shower' on Saunders Island. I was fascinated by this and resolved to see the hilarious spectacle for myself. Luckily, around 35 per cent of the total population of Southern Rockhopper Penguins breed on the Falkland Islands, so I was confident that visiting there would result in another success for Mission Penguin.

My expedition to the Antarctic Peninsula had ended in the Falkland Islands and included day visits to West Point Island and East Falkland. The extension I had booked to this trip involved staying on some other islands including my target, Saunders Island.

My first sighting of a Southern Rockhopper Penguin was on West Point Island, which I discovered was famous not only for the incredible birdlife but also the most amazing afternoon tea. The island was run as a farm and did not have guest accommodation, but welcomed visitors from expedition ships to savour its delights, both natural and gastronomic. To work up an appetite I walked the 2km (1.2 miles) uphill over rough land to the Southern Rockhopper Penguin colony on the rocky promontory known as the Devil's Nose. The excitement of what lay before me spurred me on and I covered the ground quickly, overtaking many of my fellow travellers on the way. I like to think that it was the sight of my first Southern Rockhopper Penguin that took my breath away rather than the climb to reach it! I was immediately struck by the brightness of their red eyes, which were so different to those of the other species I had seen so far.

BELOW For essentially a 'black and white bird', the Southern Rockhopper Penguin is remarkably colourful, with red eyes, orangey-red beak, pink feet and pale yellow crest feathers.

ABOVE A colony of Southern Rockhopper Penguins at 'the Rookery' with the penguin 'shower' on the far right, Saunders Island, the Falklands.

PREVIOUS PAGES Southern Rockhopper Penguins, Pebble Island, the Falklands.

As usual, I was one of the last still at the colony when our expedition crew came to round us up to return to the ship. I had to almost jog back to the settlement to partake in a quick cup of tea and delicious slice of cake. A fitting celebration of another tick for Mission Penguin, and I still had Saunders Island to come.

Saunders Island is the fourth largest of the Falkland Islands and consists of three peninsulas linked by narrow necks. The penguin 'shower' that I was so keen to witness is located at an area called 'the Rookery' and is formed by freshwater running down the side of the cliff and cascading via an overhanging rock. It was everything that I had hoped it would be, and I sat and watched for hours, giggling as the birds came up, queued in turn then took their shower. Occasional squabbles broke out if one bird was deemed to be washing for too long by those still waiting in line. It was a truly magical experience and one that I will never forget.

OPPOSITE Southern Rockhopper Penguins forming a queue for their turn in the 'shower'.

ABOVE LEFT A rather grubby Rockhopper Penguin beginning its ablutions.

ABOVE RIGHT After a good 'shower', the Southern Rockhopper Penguin looks pristine again.

RIGHT Two Southern Rockhopper Penguins squabbling at the shower as a third makes a hasty retreat.

On a subsequent visit to the Falkland Islands, I also visited Sea Lion Island. Here I watched the Southern Rockhopper Penguins make their way from the turbulent ocean up the steep cliff face to their colony. They seemed to appear out of nowhere, their heads tiny specks amidst the waves crashing against the cliffs. Suddenly they leapt out onto the steep rock face and frantically started 'hopping' their way up, desperately trying to reach the safety zone before the next wave arrived. They were frequently knocked off their feet and hurled unceremoniously back into the swirling ocean to try again. Their tenacity was incredible, and I couldn't help but wince as they bounced down off the rocks before hitting the water again.

OPPOSITE A Southern Rockhopper Penguin demonstrating the origin of its name, East Falkland.

BELOW LEFT Two Southern Rockhopper Penguins leaping out of the waves onto the near vertical cliff to reach the colony at the top, Sea Lion Island, the Falklands.

BELOW RIGHT Having successfully reached the 'safe' zone above the crashing waves, a group of Southern Rockhopper Penguins pause to preen, Sea Lion Island, the Falklands.

OPPOSITE A group of Southern Rockhopper Penguins navigating down the cliff from the colony back to the ocean, East Falkland.

ABOVE *Sequence:* A Southern Rockhopper Penguin falling as it hops and jumps its way down the steep rocky terrain, East Falkland.

Returning to the ocean could be just as hazardous, and they occasionally misjudged their jump and ended up face-first on the rocks below. It was a miracle that they didn't seem to be hurt but just pushed themselves up using their beak and flippers and carried on with their journey.

OPPOSITE A Southern Rockhopper Penguin demonstrating a perfect landing using its strong claws to grip the rocks and flippers for balance, East Falkland.

BELOW A Southern Rockhopper Penguin colony on the north coast of Port Louis, East Falkland.

It was on this visit to the Falkland Islands that I also encountered my first Southern Rockhopper Penguin chicks, as I had timed my visit to coincide with the breeding season. The chicks were already old enough to be left in crèches, guarded by a few adults, while eagerly awaiting the return of their parents with food.

BREEDING

Southern Rockhopper Penguins breed in large, dense and noisy colonies, often alongside Black-browed Albatrosses and Imperial Cormorants. These colonies can be up to 60m (197ft) above sea level. The nest is a shallow depression lined with mud, stones and feathers, and they may even make use of an abandoned Black-browed Albatross nest. The exact breeding season varies with location, but the two eggs are usually laid 4–5 days apart, in early November. The second egg is larger than the first and, after around five weeks of incubation by both parents, usually hatches first. Unless conditions are exceptionally favourable, it is this chick that is preferentially fed and survives, with the smaller egg chick dying from neglect. The chick is brooded by the male for 3–4 weeks while the female forages and provisions it. Small crèches then form, allowing both parents to forage until the chick fledges at around 10 weeks.

PREY AND PREDATORS

Southern Rockhopper Penguins eat mainly fish, squid, octopuses, krill and other small crustaceans. The eggs and chicks are predated by skuas, gulls and caracaras and at sea the adult birds are targeted by South American Sea Lions, sharks, fur seals and giant petrels.

RIGHT A Southern Rockhopper Penguin vocalising to announce its return from foraging.

OPPOSITE TOP A Southern Rockhopper Penguin standing guard over a crèche, Kidney Cove, East Falkland.

OPPOSITE BOTTOM A hungry Southern Rockhopper Penguin chick demanding food. Penguins have backward-pointing barbs in their mouth, tongue and throat to help them swallow slippery fish.

OPPOSITE The seven species of the crested penguin family usually raise only one chick. On the Falkland Islands, however, providing food is plentiful, Southern Rockhopper Penguins can raise two.

RIGHT A Southern Rockhopper Penguin chick encouraging its parent to regurgitate.

BELOW A Southern Rockhopper Penguin chick exploring under the watchful eye of a parent.

OPPOSITE *Sequence:* Southern Rockhopper Penguins returning from foraging, Saunders Island, the Falklands.

ABOVE A 'waddle' of Southern Rockhopper Penguins returning to the colony, hopefully with full stomachs to feed their hungry chicks.

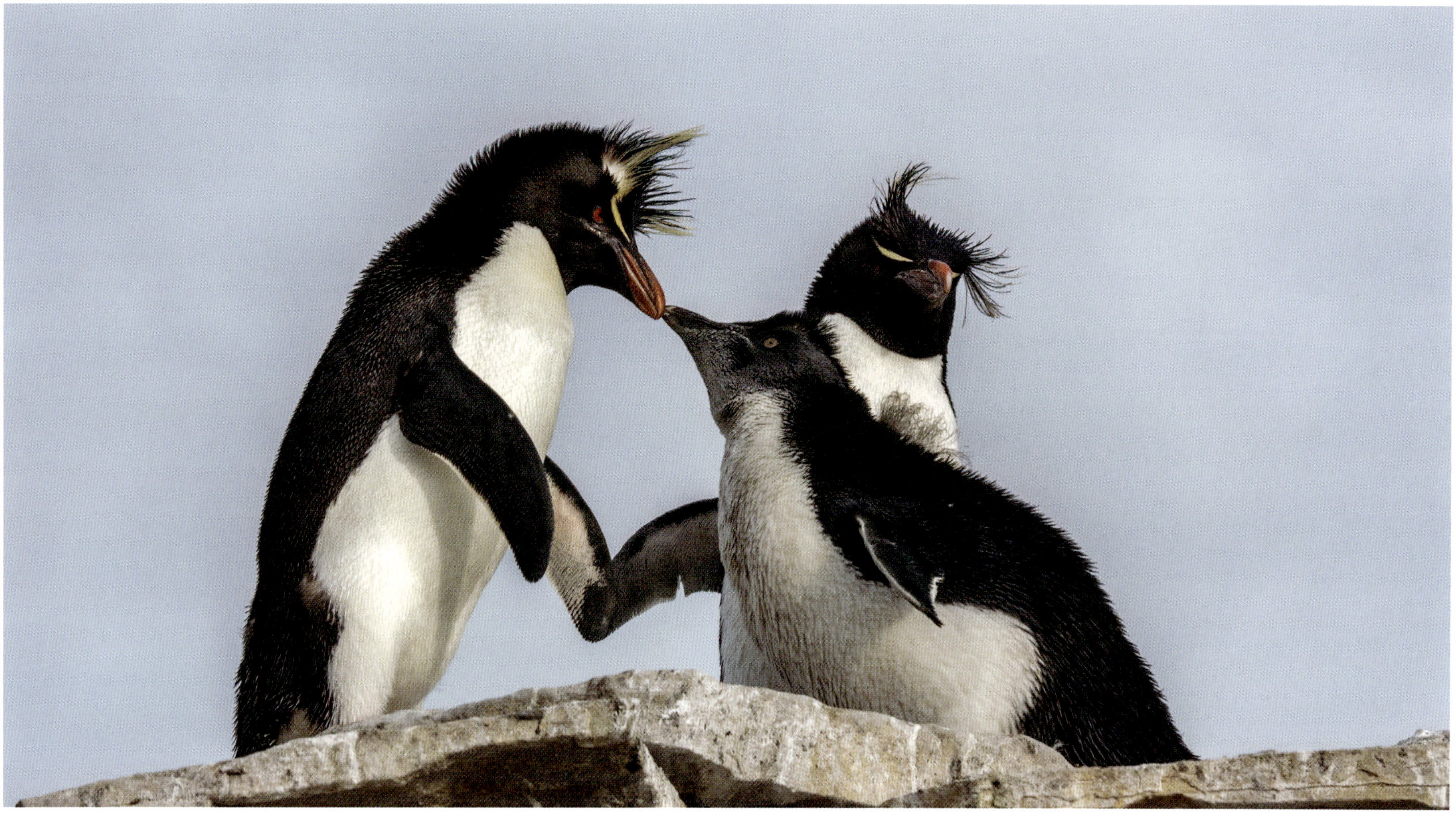
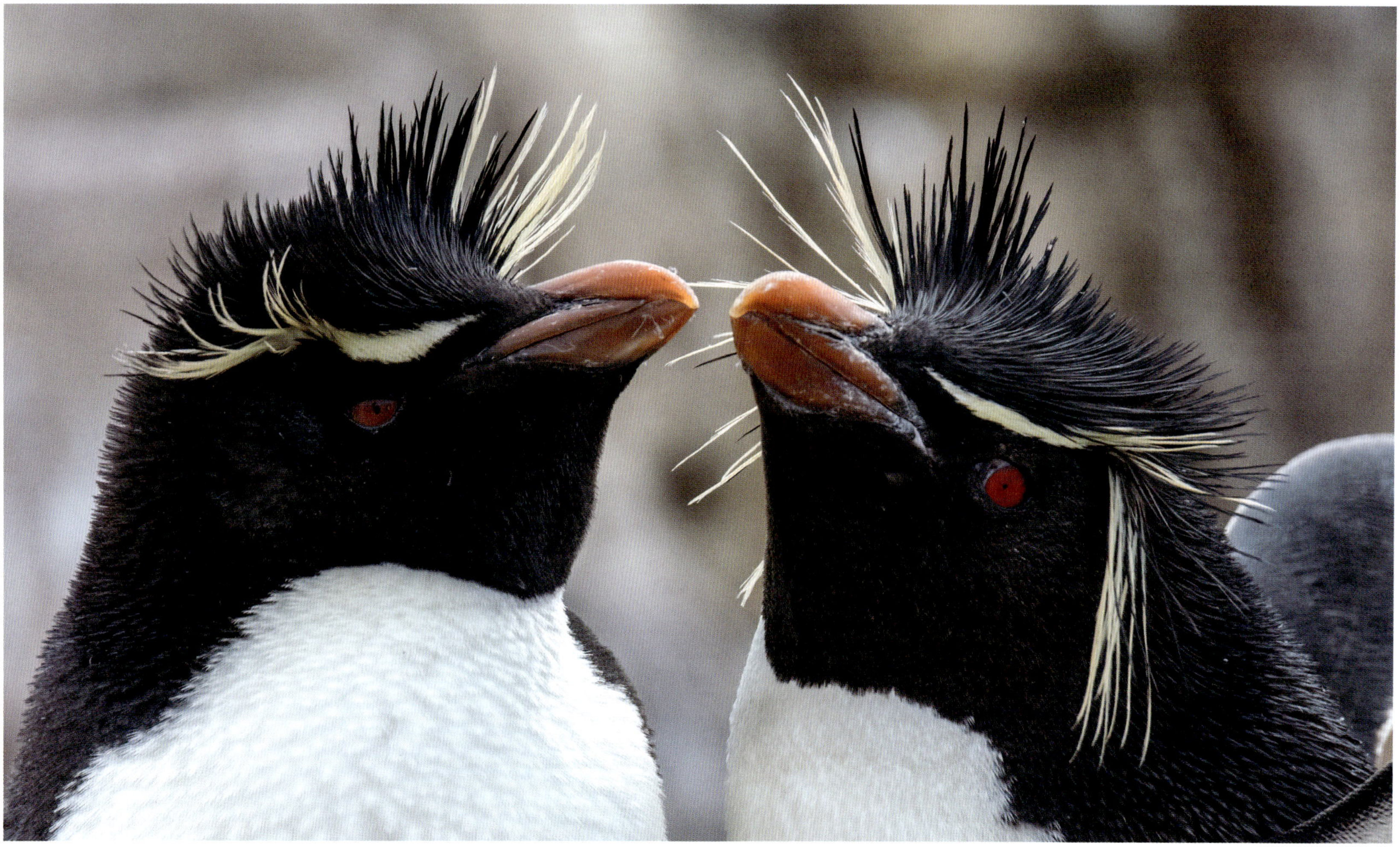

The Falkland Islands Southern Rockhopper Penguin is the western subspecies. A few years later, while on another Mission Penguin expedition, I also managed to see the eastern subspecies on Macquarie Island in the south-west Pacific Ocean. I was absolutely thrilled as it meant that, if these two subspecies are ever classified as different species, then I am one step ahead and Mission Penguin is future-proofed.

KEY THREATS

The primary threats facing Southern Rockhopper Penguins are overfishing, oil pollution, disease, climate change (affecting food quality and availability), and also toxins from algal blooms entering the food chain.

RIGHT The eastern subspecies of the Southern Rockhopper Penguin showing the differentiating bare pink skin around the beak, Macquarie Island.

OPPOSITE TOP A Southern Rockhopper Penguin chick in its immature plumage ready for fledging.

OPPOSITE BOTTOM Southern Rockhopper Penguins form lifelong pairs.

MOSELEY'S (OR NORTHERN) ROCKHOPPER PENGUIN

MOSELEY'S (OR NORTHERN) ROCKHOPPER PENGUIN	SIZE AND WEIGHT	POPULATION	CONSERVATION STATUS
Eudyptes moseleyi	55–65cm (21.7–25.6in), 1.6–4kg (3.5–8.8lb) depending on time of year and gender.	Around 413,700 mature individuals and rapidly decreasing.	Endangered (IUCN Red List 2020).

The bushy crest of the Moseley's (or Northern) Rockhopper Penguin starts as a pale yellow stripe over each bright red eye then increases significantly in length and number of plumes. It has a black back, tail, head and throat with a white front. Its feet are pink with black soles and the beak an orange-red. The smaller, duller juveniles have a very short crest; the chicks are brown and white.

Anyone who has seen the film *Happy Feet* will surely remember the guru 'Lovelace' with his amazing, wild head-crest. Well, 'Lovelace' looks like a Moseley's Rockhopper Penguin. Also known as the Northern Rockhopper Penguin, this is the most recently recognised penguin species as it used to be classed as a subspecies of the Southern Rockhopper Penguin. Although it became officially known as Moseley's Rockhopper Penguin in 2006, the name Northern Rockhopper still persists. Despite once being considered one species, these penguins' crests are very different, with Moseley's Rockhopper having the most flamboyant and longest plumes of all the crested penguins. Moseley's Rockhopper also proved to be one of the most challenging species to get to, let alone to return home from.

Approximately 90 per cent of Moseley's Rockhopper Penguins are found on the archipelago of Tristan da Cunha, a group of volcanic islands in the South Atlantic Ocean. The remaining birds breed in the Indian Ocean on Amsterdam Island and St Paul Island. Tristan da Cunha is the remotest inhabited archipelago in the world, with only around 250 resident islanders, who all live on the main island. It is about 2,816km (1,750 miles) from South Africa and just over 3,219km (2,000 miles) from South America, and there is no airstrip, so visitors arrive by boat, which is at least a six-day journey from South Africa. To put this into context, it only takes three days to reach the moon.

When I first investigated travelling to Tristan da Cunha it seemed almost impossible. There was a research and supply vessel that visited the main island of the archipelago once per year. The other suggestion was to catch a ride on a fishing trawler from Cape Town which, for someone who suffers severely from seasickness, was not an appealing option. Luckily, I finally stumbled upon an expedition with Silversea Cruises sailing from Ushuaia in South America to Cape Town in South Africa, via Tristan da Cunha. I had hoped to visit the islands during the penguins' breeding season (usually early August to early January); however, at the end of breeding the adults return to sea for a few weeks to replenish their reserves before returning to land for their annual moult. I was, therefore, hopeful that there would still be some adults on the islands in early March when the boat was scheduled to arrive. It was those timings or nothing, so along with two friends I excitedly booked a cabin.

As the 28 February 2020 departure date approached, news was building of a new coronavirus (Covid-19). There were six reported cases in the UK, but no deaths so far. We completed a health-screening questionnaire on arrival and had temperature checks before boarding the vessel, but other than that continued as planned.

Gough was the first island of the archipelago we reached, and as the ship approached I could just make out a group of black and white birds on the distant shoreline. My first Moseley's Rockhopper Penguins – I was absolutely thrilled.

BELOW Moseley's Rockhopper
Penguins, Gough Island.

PREVIOUS PAGES A group of
Moseley's Rockhopper Penguins
on Inaccessible Island.

BREEDING

Moseley's Rockhopper Penguins breed in dense, noisy colonies either on boulder-strewn beaches, rocky slopes, in crevices or in dense tussock grass. The males usually return to the colony in late July, with the females following soon after. Two eggs are laid in early September, 4–5 days apart, with the second egg being larger than the first. Both parents share incubation in shifts, and if both eggs hatch it is the chick from the second, larger egg that is raised. The smaller egg chick usually dies within the first week from starvation and neglect. The male broods the favoured chick for the first 3–4 weeks while the female provisions it. Once the chick is old enough it is left in a crèche, allowing both parents to forage until it fledges in late December.

As predicted, many of the penguins were midway through their annual moult, so looked a little scruffy to say the least. I did, however, manage to find some penguins that were newly attired in their fresh plumage and looked stunning. They were even more amazing in real life than in the film and in books, and their head-crests were just spectacular.

It was also clear to see where the breeding colony had been among the boulders and tussock grasses just above the shoreline.

OPPOSITE Moseley's Rockhopper Penguins are monogamous and form a lifelong pairing.

BELOW LEFT A Moseley's Rockhopper Penguin watching us from the rocky shoreline of Gough Island.

BELOW RIGHT The crest of this Moseley's Rockhopper Penguin blended in with the tussock grass.

Gough Island, together with Inaccessible Island, which we visited next, is a UNESCO (United Nations Educational, Scientific and Cultural Organization) World Heritage Site. It has been described as one of the least disturbed temperate islands left on Earth. Landing is, therefore, not permitted on either island, but using the Zodiacs we were able to get close to the shore and had some great sightings. Taking great photographs, however, was rather more of a challenge. The Zodiac was at the mercy of the swell and we were also buffeted by the wind, as indeed were the penguins.

We had been scheduled to land on the only inhabited island of the archipelago, Tristan da Cunha itself, and meet some of the residents. By now, however, the seriousness of Covid-19 was becoming clear and, quite rightly, the islanders refused us permission to come ashore. This meant that we had more time cruising around the other islands, so even more opportunities to see Moseley's Rockhopper Penguins. I was in heaven.

ABOVE The wind frequently catches the long crests of the Moseley's Rockhopper Penguins.

OPPOSITE A couple of Moseley's Rockhopper Penguins displaying the origins of their 'rockhopper' name.

PREY AND PREDATORS

Moseley's Rockhopper Penguins eat mainly krill, squid, octopuses and some small fish. The chicks and eggs are predated by Subantarctic Skuas, while the adults are preyed upon by giant petrels, sharks and Subantarctic Fur Seals.

OPPPOSITE A Greater-crested Tern returning to its nest after a fishing trip, Cape Town harbour.

BELOW A Moseley's Rockhopper Penguin carefully navigating its way through the Subantarctic Fur Seals.

Our ship then continued towards Cape Town as planned, and with each nautical mile travelled the news of the coronavirus worsened. By the time we could sight land, all the ports in South Africa had been closed and we were stuck out on the ocean just outside Cape Town – so near and yet so far. The captain finally managed to negotiate our way in the following day, albeit initially to the container port, and we were not allowed to lower the gangway. Never one to miss an opportunity, I spent most of my time out on deck photographing any wildlife that swam or flew by, including Greater-crested Terns.

We were told that we would probably be held on the ship for several days. However, the following day approximately 20 of us had our names called, were told to pack and were escorted straight to the airport for a flight to London Heathrow. At the desk our passports were stamped for both entry into South Africa and exit. The officer joked that it must be the shortest holiday on record.

We arrived home in the UK to find bare shelves in our local village store, and were then straight into the first UK lockdown for Covid-19. What a huge relief to have got home safely when we did, and what wonderful memories to sustain me through lockdown as I sorted the myriad of photographs of Moseley's Rockhopper Penguins.

FIORDLAND PENGUIN

FIORDLAND PENGUIN *Eudyptes pachyrhynchus*	SIZE AND WEIGHT 55–60cm (21.7–23.6in), 2.1–5.1kg (4.6–11.2lb) depending on time of year and gender.	POPULATION 12,500–50,000 mature individuals and decreasing.	CONSERVATION STATUS Near Threatened (IUCN Red List 2020).

The crest of the Fiordland Penguin is a broad pale yellow stripe over each eye, with slightly longer plumes towards the back of the head. It has a white front, bluish-black back and tail, and black head and throat with a few white lines across its cheeks. It has pink feet with black soles and an orange-red beak. The duller juveniles have a shorter crest; the chicks are dark brown and white.

After my husband died, I made the decision to travel to places that we had not previously visited together. This was my way of moving forward and helping to ensure that I didn't try to live my life in the past. Before we were married, however, Ralph had lived and worked in New Zealand for six years, so inevitably during our marriage we had returned for several wonderful holidays. Whereas we had seen the Little Penguin and the Yellow-eyed Penguin while there together, we had not trekked into the forest to see the endemic Fiordland Penguin. To complete Mission Penguin, therefore, I needed to retrace our steps to South Island, New Zealand. I must confess that I was full of trepidation as I left the UK, and my anxiety continued to build during the journey. I need not have worried. It was as if the country knew of my pain and embraced me in its warm familiarity. I was instantly soothed, and the excitement of going in search of the Fiordland Penguin took over. The penguins are found predominantly on the west to south-west coast of South Island and are one of the few penguin species that nest in mature temperate rainforest.

RIGHT Fiordland Penguins at the edge of the dense temperate rainforest, clearly showing the white face markings that distinguish them from the other crested penguin species.

PREVIOUS PAGES Fiordland Penguins on the beach, South Island, New Zealand.

BREEDING

Fiordland Penguins breed in loose colonies in a range of habitats including caves, rocky shorelines and temperate rainforest. The males usually return to the colony in June, with the females following soon after. Two eggs are laid about four days apart in a shallow scrape or hollow, with the second egg being larger than the first and hatching first. It is generally this chick that is raised, with the smaller egg chick usually being neglected and dying from starvation. The favoured chick is brooded by the male and provisioned by the female for the first few weeks. It then joins a small crèche, allowing both parents to forage, and fledges at around 11 weeks.

The colony that I had selected to visit was on the Lake Moeraki coastline. It was in a protected area so only accessible in very small groups with a licensed conservation guide. The walk down through the rainforest was certainly not for the faint-hearted or the unfit. From the roadside above we entered the dense forest and clambered down the muddy slope, grabbing at tree trunks and flimsy branches to both slow the descent and maintain our balance. We then followed the path of a small stream, which we had to cross several times as it meandered back and forth along our way. There were no bridges, and the recent rains had turned the stream into more of a river, which was at times deeper than our boots. The fast-flowing current also necessitated us forming human chains to help each other across safely. Wet feet did not deter me, so long as I could keep my camera dry. We continued to slip and slide our way towards the beach, stepping over fallen trees and detouring around impassable clumps of vegetation. The penguins take all this in their stride and use their strong claws to grip the slippery terrain. They were probably more amused at our antics that day than we were at theirs.

Finally, we arrived at the beach and made our way to a rocky area where we could sit quietly and watch without disturbing the penguins. There was a constant flow of penguins waddling past us in both directions as they made their way to and from the colony. There was certainly no mistaking who was going where. The birds coming from the Tasman Sea to return to their chicks were pristine, whereas the penguins leaving the forest colony were as muddy as we were.

BELOW LEFT A rather muddy Fiordland Penguin emerging from the forest to go foraging.

BELOW RIGHT A pristine Fiordland Penguin returning from the Tasman Sea.

OPPOSITE A clean Fiordland Penguin returns from the sea, passing a dirty one heading from the colony.

I had a wonderful and mesmerising two hours watching and photographing the penguins. It was fascinating observing the different routes they took home, with some attempting to scale sheer rock faces, as it was a shorter distance, and others taking the 'penguin path', which was longer but invariably quicker. The fastest route back to the sea was unanimously jumping down the stony slopes.

PREY AND PREDATORS

Fiordland Penguins eat mainly squid, krill and small fish. At sea the adults are predated mainly by New Zealand (Hooker's) Sea Lions, while on land introduced predators such as stoats, dogs, ferrets, rats and feral cats can decimate colonies.

ABOVE The penguins have a well-used path from the sea back up to the colony.

OPPOSITE A Fiordland Penguin using its strong claws to grip the rock face on a shortcut back up to the colony.

RIGHT Squabbles often break
out on the narrow paths.

OPPOSITE The quickest way
down from the forest to the
beach is to jump!

Mission Fiordland Penguin
was accomplished, other than the
not insignificant task of retracing
our steps back out of the forest.
The ascent up the muddy slope
proved just as challenging as the
descent but, inspired by the intrepid
penguins, I made it out too.

KEY THREATS

Commercial fishing, especially
for squid, is a serious threat to
Fiordland Penguins, both in terms
of competition for prey and as
bycatch. Pollution and oil spills
are also significant threats.

SNARES PENGUIN

SNARES PENGUIN *Eudyptes robustus*	SIZE AND WEIGHT 51–61cm (20.1–24in), 2.4–4.3kg (5.3–9.5lb) depending on time of year and gender.	POPULATION Around 63,000 mature individuals and stable.	CONSERVATION STATUS Vulnerable (IUCN Red List 2018).

The Snares Penguin's crest is a broad pale yellow stripe that curves over each eye with slightly longer plumes towards the back of the head. It has a bluish-black back and tail, a white front, a black head and face, a large orange-red beak and pink feet with black soles. The smaller juveniles have a grey throat and a shorter, paler crest; the chicks are dark grey and white.

Before embarking on Mission Penguin, I had never even heard of the Snares Islands, let alone the endemic Snares Penguin that breeds there. The small group of remote islands are approximately 200km (124 miles) south of New Zealand's South Island and are certainly not on any list of popular tourist destinations. After a lot of research, I finally discovered a three-week boat trip called 'Birding Down Under', run by Heritage Expeditions on MV *Spirit of Enderby* (formerly called MV *Professor Khromov*). This small Russian ship was built in 1983 for research and later refurbished to hold 50 passengers. One key thing about this ship, which I did not discover until after embarking, was that it had no external stabilisers. The Snares Islands occupy the tempestuous latitudes of the Southern Ocean known as the 'Roaring Forties'. The ship therefore not only went up and down with the waves but also rocked side to side. At times it felt like being inside a washing machine, and even managing to stay in bed at night without rolling out required a self-made barrier of pillows and fleeces. I also suffer badly from seasickness, even on large ferries and flat seas. Suffice to say I felt very unwell at times. Even that, however, could not detract from the excitement and awe of this incredible adventure, which became one of my favourite expeditions.

The islands were appropriately called 'the Snares' as they 'snare' ships. This meant that we could not get in close, and the only way to see the islands was to use Zodiacs. On the way we were informed by the expedition leader that for the last few years the sea conditions had been too rough to launch them, so to keep our fingers crossed. I must confess that I had everything crossed and said a little prayer. While I could probably have spotted a Snares Penguin in the far distance with my binoculars and therefore have a tick for Mission Penguin, I really wanted to see them up close. I cannot tell you how thankful and relieved I was to hear the 'go' decision later that day.

Since the pristine Snares Islands are part of the New Zealand Subantarctic Islands UNESCO World Heritage Site, we were not allowed to land, but the Zodiacs could cruise close to the shoreline. As we approached the main island of the group, North East Island, I could see the beautiful forests of the large tree daisy (*Olearia lyallii*). I was particularly excited as it is within these forests that the Snares Penguins gather in groups to breed. As we drew closer, I could hear the raucous calls of the birds and then, as we rounded the bend on the eastern side of the island, there they were.

BELOW About 80 per cent of the Snares North East Island is covered by forests of the large tree daisy (*Olearia lyallii*), forming a canopy over 5m (16ft 5in) tall in places.

PREVIOUS PAGES Snares Penguins, Snares Islands.

ABOVE Snares Penguins returning from the ocean.

OPPOSITE Snares Penguins have a high pair-bond fidelity season to season.

The shoreline was covered in bull kelp, which surged with every wave, creating a moving surface for the penguins to negotiate as they departed from, or returned to, the island. The birds would stumble, slip, slide and frequently fall back into the water before trying again. Our Zodiac also rolled in the waves and I have never taken so many photographs of empty sea or sky. Luckily, I did manage to time a few just right and so came away with some shots that did contain penguins.

BREEDING

Snares Penguins nest in dense colonies up to 70m (230ft) above sea level under the forest canopy. The nests are usually made from twigs, stones and mud, but some individuals nest in natural cavities. The males return to the colony around September, with the females following about a week later. Two eggs are laid about four days apart, with the second egg being larger than the first. The eggs are incubated by both parents, and although both eggs hatch, only one chick is raised. This is usually from the larger egg, with the smaller chick starving through neglect. The favoured chick is provisioned by the female for the first few weeks and brooded by the male. It then joins a small crèche, allowing both parents to forage, and it fledges after 10–11 weeks.

ABOVE There are over 100 species of kelp and seaweed on the Snares Islands for the Snares Penguins to navigate as they travel to and from the ocean.

RIGHT The 'penguin slide' on North East Island.

OPPOSITE TOP Snares Penguins queuing to leave the island via the 'penguin slide'.

OPPOSITE BOTTOM Snares Penguins propelling themselves back onto the island for the long climb back up to the forest.

On the eastern shore, between all the undulating bull kelp, there is a very small area of exposed, smooth bedrock known as the 'penguin slide'. This is a key entry and exit point for the penguins as it avoids the surging bull kelp, but timing is everything, especially for birds returning to the island. By releasing trapped air from under their feathers they can propel themselves out of the water, giving them a better chance of making it beyond the splash zone before the next wave comes to drag them back down. Not all make it first time, however, and it was a wonderful demonstration of perseverance. It was also a rather amusing spectacle that will stay with me forever. For the penguins, of course, it was more serious, as the lives of their chicks depended upon their safe return from foraging. I felt so lucky that we had been able to launch the Zodiacs and so privileged to have spent time watching these engaging and tenacious creatures.

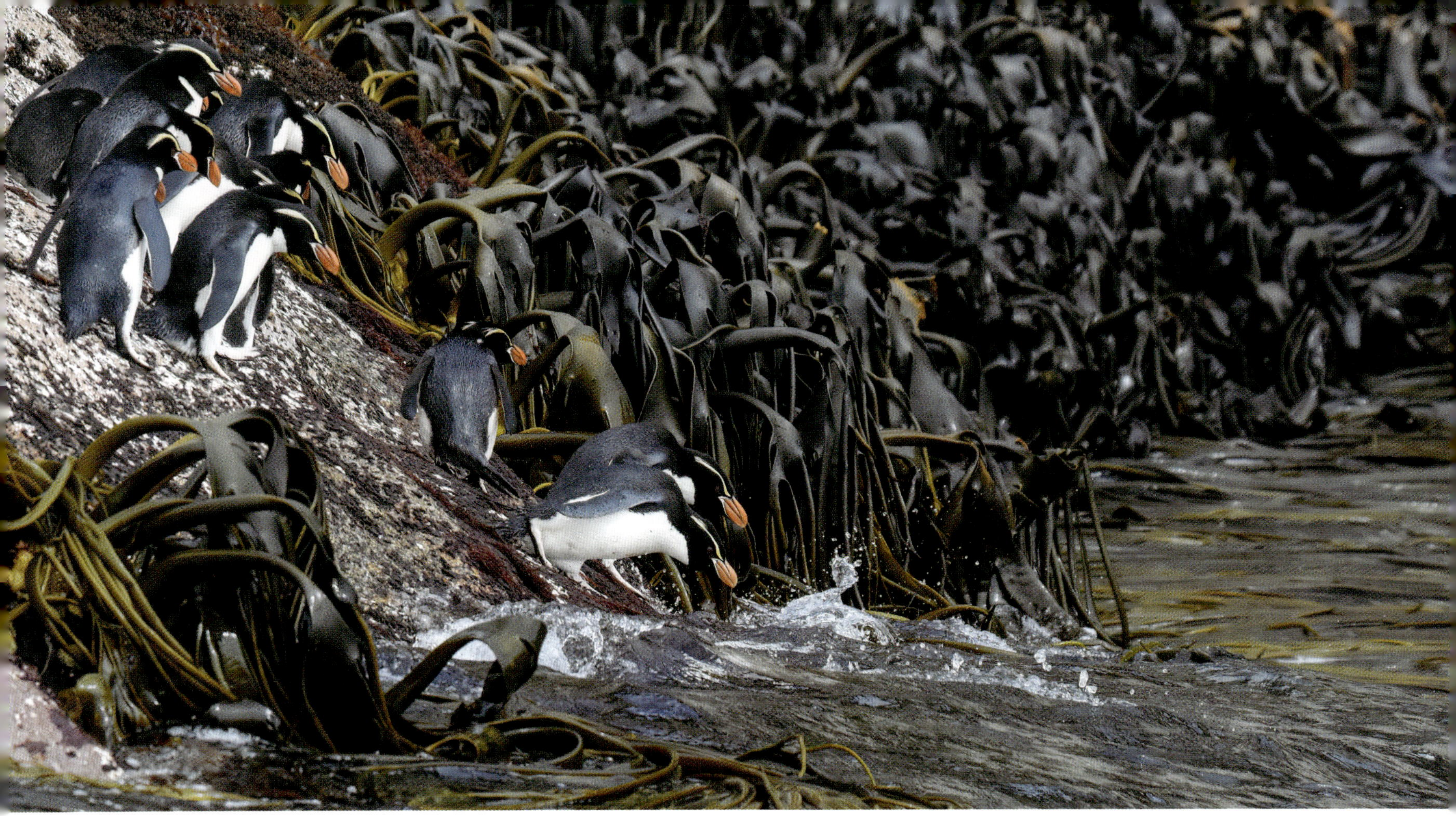

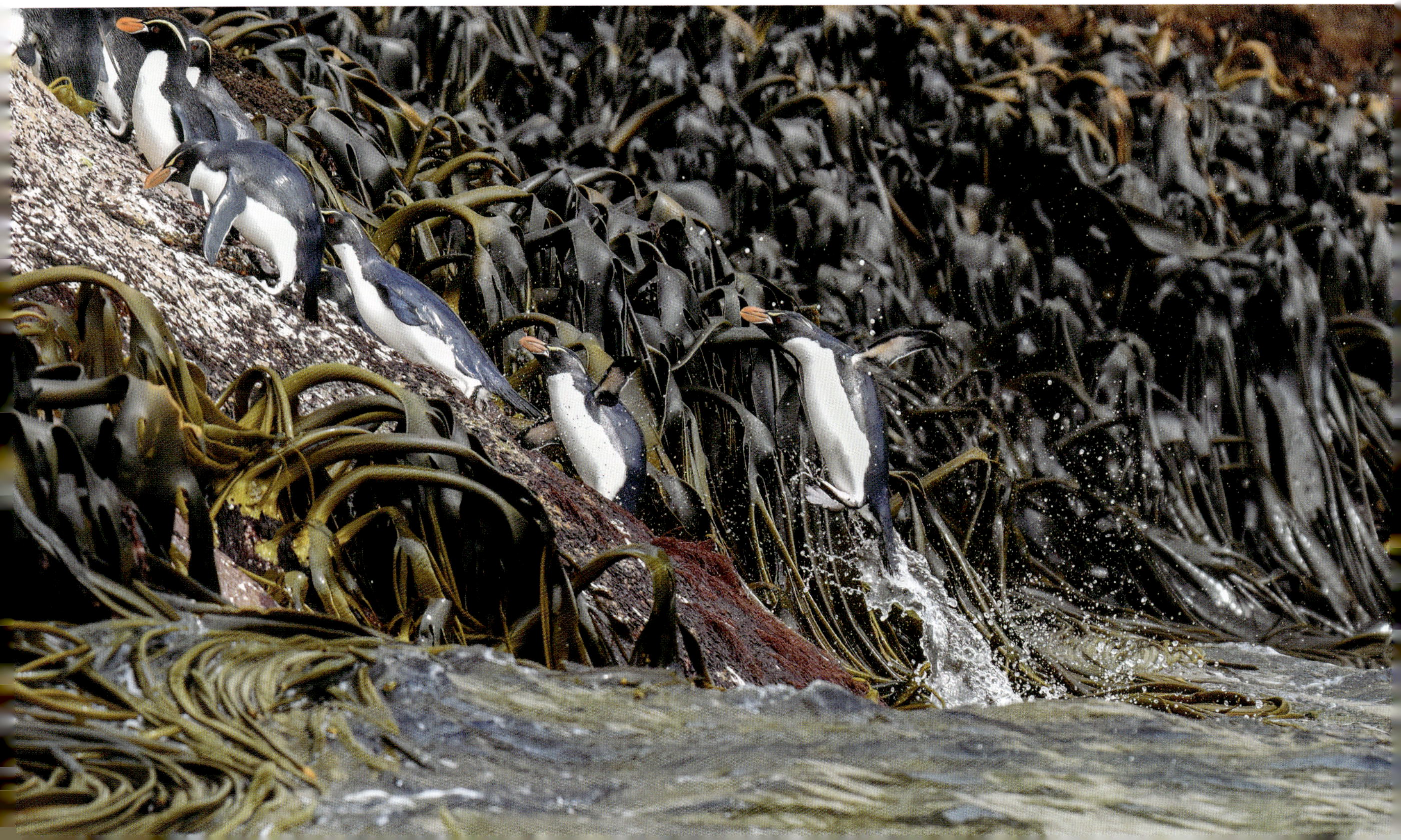

PREY AND PREDATORS

Snares Penguins eat predominantly krill, squid and small fish. The eggs and chicks are predated by skuas while the adult birds can be targeted by New Zealand Sea Lions and occasionally Leopard Seals, although these are not common around the Snares Islands.

TOP LEFT The buttery-yellow bushy crest of the Snares Penguin starts just above the beak and curves slightly over the eyes before flaring outwards towards the back of the head.

TOP RIGHT When wet, the crests flatten to a thin stripe over the eyes.

OPPOSITE Snares Penguins use their strong gripping claws to clamber up the slippery cliff face to feed their chicks eagerly waiting in the forest above.

KEY THREATS

The very narrow breeding range of the Snares Penguin makes it vulnerable to sudden unpredictable events, such as oil spills. Commercial fishing and the effects of climate change on its food supply are the most likely long-term threats.

ROYAL PENGUIN

ROYAL PENGUIN	SIZE AND WEIGHT	POPULATION	CONSERVATION STATUS
Eudyptes schlegeli	65–75cm (25.6–29.5in), 3–8.1kg (6.6–17.9lb) depending on time of year and gender.	Around 1.5 million mature individuals (population trend unknown).	Least Concern (IUCN Red List 2022).

Starting at the forehead and flaring back over the eyes, the Royal Penguin's head-crest is orange-yellow streaked with black. It has a bluish-black back, tail and crown, a white front, a whitish face and throat and pink feet with black soles. The orange-brown beak has some visible skin at the base and a pale yellow tinge below. The duller juveniles lack the full crest; the chicks are brown and white.

Macquarie Island lies in the south-west Pacific Ocean, about halfway between New Zealand and Antarctica. At a latitude of 54°S it is right in the path of the infamous 'Furious Fifties' winds, so not the best location for someone who suffers from seasickness. However, as 99.9 per cent of the world population of Royal Penguins breed on Macquarie Island it was an essential destination for Mission Penguin. Luckily this island, although Australian, was included in the itinerary of my expedition to the New Zealand Subantarctic Islands, so did not require extensive further planning to visit other than an Australian visa. Although there are no permanent inhabitants, the Australian Antarctic Division has a base at the far north of the island, and we were permitted to land. I cannot tell you how wonderful it was to feel the ground beneath my feet again.

The beach was full of Southern Elephant Seals, including pups, moulting adults and enormous males, battle-weary and scarred from their many fights to either gain or maintain their harem of females. As I approached the base, a gorgeous smell of freshly baked scones filled the air. Tempting as it was, it still took me a little while to partake as I had to pass several Gentoo Penguins en route that were just asking to be photographed. The staff at the base were so welcoming and hopefully enjoyed our visit as much as we did. Apparently, these subantarctic islands are less visited than Antarctica, so they don't see many new faces.

After some delicious sustenance I explored the north of the island, including the rusted remains of the digesters that once consumed up to 4,000 penguins per day. The animals were slaughtered for their oil, which was used for fuel, lighting and for tanning leather. Penguin skin was also in demand for handbags, hats and slippers, and their feathers were used to fill pillows and mattresses. The two main penguin species of

ABOVE The rusted remains of the digesters on Macquarie Island, where seals and then penguins were slaughtered for their oil in the 1800s.

Macquarie Island, the King Penguin and Royal Penguin, were both targeted, and by the time an international campaign put an end to the carnage in 1919, millions of penguins had been killed, leaving just a few thousand remaining. As I tearfully gazed on the rusted remains, overwhelmed by the cruelty of humankind, a family of Gentoo Penguins waddled over to greet me and soon restored my smile.

BELOW These Royal Penguins certainly look regal, although the origin of their name is still unclear. Some say it is because their bright crest resembles a crown, others because they were found alongside the King Penguin and the theme was continued.

PREVIOUS PAGES Royal Penguins, Sandy Bay, Macquarie Island.

I had to wait until the afternoon and our landing at Sandy Bay before I had my first sighting of the endemic Royal Penguin I had travelled so far to see and photograph. My one concern was the potential weather, as Macquarie Island is known as an island of wind and rain. While I believe that there is no such thing as bad weather, just the wrong clothes, rain and wind do make photography incredibly challenging, especially keeping equipment dry and the camera steady. Someone was certainly looking after me that afternoon as the winds dropped and the clouds dispersed just as the Royal Penguins came in from the surf and waddled across the beach towards me.

I watched their hilarious antics and their interactions with each other, as well as their obvious curiosity of me. As they came closer to investigate and even pecked at my boots, I marvelled at how trusting they were of humans despite a century of persecution. What a privilege to share their world, if only for an afternoon.

ABOVE Royal Penguins waddling across Sandy Bay, Macquarie Island.

I had seen Macaroni Penguins before, with their long, bright yellow, orange and black plumes that meet in the middle of their forehead. It was instantly apparent from their appearance that the Royal Penguin is closely related to the Macaroni Penguin; in fact, some say that the Royal Penguin is a subspecies of the Macaroni. The only obvious difference between the two is the colour of their cheeks and throat, which are pure white to pale grey on the Royal Penguin and jet black to dark grey on the Macaroni Penguin. As I looked at the sea of white faces before me, one of our expedition guides pointed out a single stray Macaroni Penguin among them. It was as if the Macaroni Penguin knew we needed a comparison.

BELOW The Royal Penguin's white cheeks distinguish it from the similar-crested Macaroni Penguin.

RIGHT Macaroni Penguin.

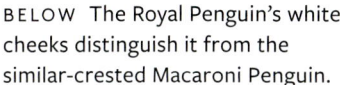

Having seen the penguins on the beach, I made my way up the muddy terrain to reach their breeding colony, which was in a huge scrape surrounded by tussock grass. There are no trees on Macquarie Island, so the tall tussock grasses provide valuable shelter from the ferocious winds.

ABOVE Two Royal Penguins squabbling in the surf.

LEFT Royal Penguins breed in dense colonies with every space taken.

BREEDING

Royal Penguins usually form a long-term pair bond and return to the same nest site each year. The males return to the colony in mid-September, with the females following in early October. The dense, noisy colonies can be up to 200m (656ft) above sea level and around 1.5km (0.9 miles) inland, and are accessed via muddy creeks. Royal Penguins raise one chick, although two eggs are laid about four days apart. The female usually displaces the first egg before the second, larger egg is laid, and it is this second egg that usually hatches. The male broods the chick for the first few weeks, with the female provisioning it. After 3–4 weeks the chick joins a crèche, allowing both parents to forage until it fledges at 9–10 weeks.

I scoured the colony with my binoculars, hoping to see an early chick, but at least found a pair proudly showing off their first egg. I wasn't the only one taking such a keen interest in exposed eggs and chicks, as large numbers of Great Skuas were constantly circling overhead. I decided I wanted to capture the intensity of this danger in a photograph, which proved rather more challenging than I initially thought. Eventually, by switching to manual focus and setting a distance just in front of me I was finally able to catch a Great Skua as it flew over my head so that it dominated the image.

BELOW A Royal Penguin pair
with an egg.

The Royal Penguin is known as one of the more argumentative penguin species, and as I watched them, squabbles were breaking out everywhere, with birds being pecked and jostled at every turn. This made travelling to and from the colony for foraging rather challenging.

PREY AND PREDATORS

Royal Penguins eat mainly krill, but also some other small crustaceans, squid and small fish. The eggs and chicks are predated by Great Skuas and giant petrels while the adult birds are targeted mainly by fur seals.

BELOW A Great Skua circling the colony, on the lookout for any unattended eggs or chicks.

RIGHT This Royal Penguin is pecked and harried as it makes its way through the colony.

BELOW Navigating their way across the beach, Royal Penguins have to move carefully between the enormous Southern Elephant Seals to avoid being crushed to death.

Sadly, it was time to tear myself away and return to the beach for the short journey back to the ship. This meant navigating the huge Southern Elephant Seals again, and it was easy to see that one of the main dangers facing these penguins is not being predated but accidentally crushed to death. Despite this parting thought, it had been a truly wonderful day and one more success for Mission Penguin.

KEY THREATS

The effect of climate change on their food supply is the most likely long-term threat facing Royal Penguins. In addition, their very narrow breeding range makes them vulnerable to sudden unpredictable events, such as oil spills or natural disasters.

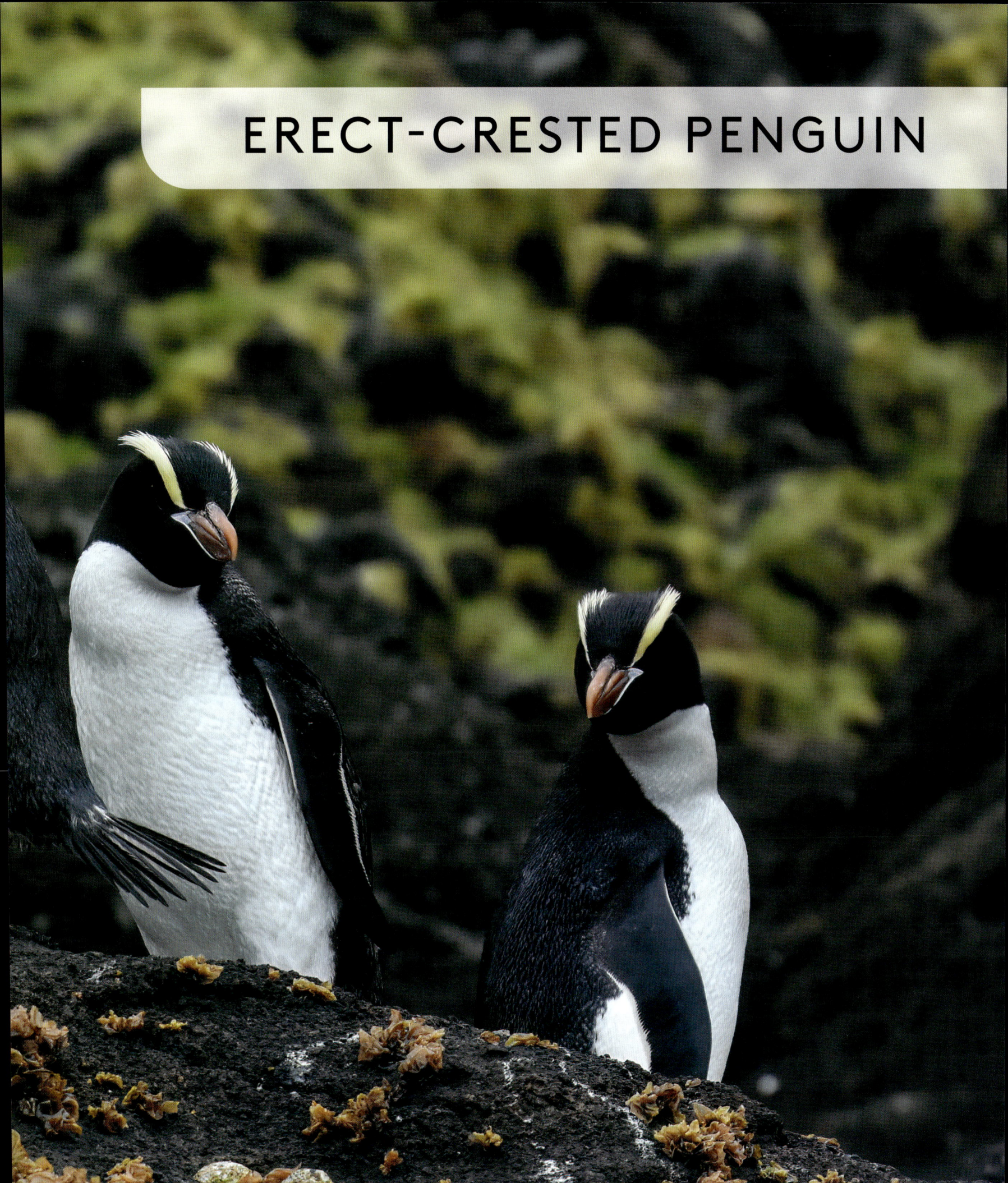

ERECT-CRESTED PENGUIN

ERECT-CRESTED PENGUIN *Eudyptes sclateri*	SIZE AND WEIGHT	POPULATION	CONSERVATION STATUS
	60–67cm (23.6–26.4in), 3–7kg (6.6–15.4lb) depending on time of year and gender.	Around 150,000 mature individuals and decreasing.	Endangered (IUCN Red List 2020).

The only crested penguin that can voluntarily raise and lower its crest, the Erect-crested Penguin has a bushy pale yellow crest starting at the top of its orange-red beak and extending back over each eye. It has a black back, tail, head and throat, white front and pink feet with black soles. The duller juveniles have a shorter, paler, non-erectile crest; the chicks are brown and white.

The Erect-crested Penguin breeds on the Antipodes Islands and Bounty Islands, which are part of the New Zealand Subantarctic Islands UNESCO World Heritage Site. Both are remote, uninhabited volcanic archipelagos which are inhospitable to humans but fortunately not to penguins. Luckily, the 'Birding Down Under' expedition included both locations on its itinerary, which doubled my chances of seeing this remote crested penguin.

We arrived at the Antipodes Islands in the morning to strong winds and a 4m swell. The combination made it too dangerous to launch the Zodiacs and we spent the morning circumnavigating the archipelago in search of shelter. With my binoculars I could just make out the black and white birds, although they were so distant that it was impossible to definitively identify them other than by a very strong likelihood based on our location.

Our expedition leader was determined to maximise our experience by making every possible proposed Zodiac excursion on the itinerary a reality. I knew, therefore, that if there was any chance of getting us off the ship to get closer to the penguins, he would take it. After an anxious morning waiting, the call finally came in the early afternoon, and we set off. Photography was inevitably difficult in the waves, but I managed to take a few photographs I was pleased with, and we still had the Bounty Islands to come.

ABOVE Penguins use their incredibly strong claws to climb up slippery near-vertical rock faces. The immature bird on the right lacks the erect-crest and has a grey throat.

OPPOSITE Erect-crested Penguin colony at South Bay on the main Antipodes Island.

RIGHT An Erect-crested Penguin clearly showing the origins of its name.

PREVIOUS PAGES Erect-crested Penguins, Antipodes Island.

The Bounty Islands were discovered by Captain William Bligh in 1788, and he named them after his ship HMS *Bounty*, just a few months before the infamous mutiny. Bligh recorded 'white spots like patches of snow', which we now know to be guano deposited by generations of seabirds, including the Erect-crested Penguin. I am sure that if Captain Bligh had got close enough to smell the white patches, he would have realised his error.

When we approached the Bounty Islands they were shrouded in mist and the swell was even worse than the day before. It was again looking doubtful whether we would be able to launch the Zodiacs, especially as the sea level between the gangway and Zodiac oscillated by around 2m. As it eased slightly, the burly crew helped hugely in manhandling those of us brave enough to chance it. I was lucky and kept myself, and more importantly my camera, dry. Some others were not quite as fortunate. It was worth taking the chance, as the mist eventually lifted and the penguins were finally revealed.

Unlike on the Antipodes archipelago, there is no greenery on the Bounty Islands. The total area of the small group of low granite islets is just under 1.4 square kilometres (0.5 square miles), and during the breeding season every available space is taken up with penguins nesting among colonies of Salvin's Albatross and various prion species.

OPPOSITE Erect-crested Penguins with Salvin's Albatross on the Bounty Islands.

BELOW Erect-crested Penguins on a granite ledge just above the shoreline, the Bounty Islands.

BREEDING

Erect-crested Penguins breed in dense, noisy colonies of many thousands of pairs on rocky terrain from just above the splash zone to 75m (246ft) above sea level. The males return to the colony in early September, with the females following about two weeks later. Like other crested penguins, the Erect-crested Penguin lays two eggs but only rears one chick. The eggs are laid about four days apart, with the second egg being very much larger than the first. The eggs are incubated by both parents for five weeks, with the first egg usually being lost within the first four days. The chick is provisioned by the female for the first few weeks while being brooded by the male. It then joins a small crèche, allowing both parents to forage, and fledges after about 10 weeks.

As the penguins returned to the islands from their foraging trips, they porpoised right past the Zodiacs. What a magical experience, and although it was very challenging to photograph in the constant swell I was thrilled to manage a few dramatic action shots. Who said that penguins can't fly. I too was flying high after another superb and successful day for Mission Penguin.

PREY AND PREDATORS

Erect-crested Penguins are thought to eat mainly krill, other crustaceans and squid. The eggs and chicks are predated by skuas while the adult birds are occasionally targeted by fur seals.

LEFT An Erect-crested Penguin leaping through the waves.

ABOVE An Erect-crested Penguin porpoising on its return to the islands. This technique enables them to breathe while not losing momentum as they 'fly' through the water.

KEY THREATS

The most likely long-term threat facing Erect-crested Penguins is the effect of climate change on their food supply.

Meet the
SHY PENGUIN
Megadyptes

and the
NOCTURNAL PENGUIN
Eudyptula

Whereas 16 of the 18 penguin species fall into four family groups (genera) with multiple members in each, the Yellow-eyed Penguin and the Little Penguin are both the only species in their respective genera.

The Little Penguin is, as the name suggests, the smallest of all the penguin species, and it is also the only nocturnal penguin on land.

The Yellow-eyed Penguin is not truly nocturnal like the Little Penguin; however, it does often commute to and from its nest at dawn and dusk and is the shyest of all the penguin species.

Yellow-eyed Penguin

Little Penguin

YELLOW-EYED PENGUIN

YELLOW-EYED PENGUIN	SIZE AND WEIGHT	POPULATION	CONSERVATION STATUS
Megadyptes antipodes	56–78cm (22–30.7in), 3.6–8.9kg (7.9–19.6lb) depending on time of year and gender.	2,600–3,000 mature individuals and decreasing (especially the northern population on South Island, New Zealand).	Endangered (IUCN Red List 2020).

The Yellow-eyed Penguin has a pale yellow head streaked with black, a slightly darker yellow band around the crown and pale yellow eyes. It has a dark bluish-grey back and tail and a white front. The beak is orange-red and pink, and the feet are pink with black soles. The juveniles lack the yellow headband and have a whitish throat; the chicks are dark brown.

Before my husband died and I started Mission Penguin, I had already seen three penguin species, and one of these was the very shy and rare Yellow-eyed Penguin. We had seen one in the far distance walking up the beach from the ocean on the Otago Peninsula, South Island, New Zealand. I did have a poor photograph, so technically I had both seen and photographed this species and it was, therefore, not a new target for Mission Penguin.

The Yellow-eyed Penguin is only found in New Zealand, but, as well as South Island, this includes the offshore Stewart Island and the subantarctic Auckland and Campbell island archipelagos. These latter two island groups were included on the 'Birding Down Under' expedition and each has a wealth of unique flora and fauna. They are particularly famed for their extensive areas of very large wildflowers, called megaherbs, which have evolved huge

LEFT My original distant photograph of a Yellow-eyed Penguin on a beach on South Island, New Zealand.

PREVIOUS PAGES A Yellow-eyed Penguin, Enderby Island.

RIGHT The Yellow-eyed Penguins
are not as social as other penguins,
so are usually seen alone.

leaves to adapt to the harsh conditions on the islands. As our Russian-built ship had been renamed the MV *Spirit of Enderby,* it was wonderful to visit the small but stunning Enderby Island. This island is probably the most beautiful of all the Auckland Islands and it is also home to one of the largest breeding colonies of Yellow-eyed Penguins – how lucky was I.

BREEDING

Yellow-eyed Penguins have a relatively long breeding cycle that begins with courtship activity in mid-August and continues until the chicks fledge around March. There is high pair-bond fidelity during their lifetime and both birds build the shallow nest of twigs, leaves and grass. They nest in a variety of habitats including coastal forests, sand dunes, scrub and pasture, and prefer privacy, so usually nest out of sight of other pairs. Yellow-eyed Penguins lay two eggs between mid-September and mid-October, and incubation is shared by both parents. The eggs are laid a few days apart but tend to hatch on the same day in early November, and both chicks are raised. Both parents share the parenting duties and, unlike in many species, the chicks do not form a crèche. They fledge at around 15 weeks.

We were there predominantly to see the amazing megaherbs, endemic birds (including the Auckland Snipe, Shag and flightless Teal), two species of albatross and the rare New Zealand Sea Lion. While these were all wonderful, my excitement grew exponentially at the sight of one, then two, then three Yellow-eyed Penguins in the distance. They were well dispersed, which fits with their reputation as the shyest and most private of all the penguin species, but we were able to get relatively close without disturbing them. With my binoculars and then long camera lens it was easy to see where the name 'yellow-eyed' came from. They also, however, have pale yellow heads with black feather shafts and a slightly brighter yellow band from their eyes around the back of their head, features that had not been possible for me to see from my original distant photograph. I was thrilled that despite the inclement weather, including sleet, I was able to take some much better photographs more befitting of Mission Penguin.

PREY AND PREDATORS

Yellow-eyed Penguins tend to forage along the seafloor and eat a variety of fish including sprats, Silversides and Blue Cod. In some areas, especially on the populated South Island, the eggs and chicks are predated by feral cats, stoats and ferrets. The adult birds can also be injured or killed on land by dog attacks, while at sea they are predated by New Zealand Sea Lions and sharks.

On our way back to the ship from the island we even saw a few more adult and juvenile penguins scattered along the dark shoreline, perhaps about to go foraging. It had been an amazing day, and although I had not set out specifically to see the Yellow-eyed Penguin I was as overjoyed as if it had been a new species for me.

OPPOSITE A Yellow-eyed Penguin seemingly unperturbed by the light sleet on Enderby Island.

RIGHT Immature Yellow-eyed Penguins have a greyer head and lack the yellow-coloured head band.

LITTLE PENGUIN	SIZE AND WEIGHT	POPULATION	CONSERVATION STATUS
Eudyptula minor	30–40cm (11.8–15.7in), 1–1.5kg (2.2–3.3lb) depending on time of year and gender.	Around 469,760 mature individuals and stable.	Least Concern (IUCN Red List 2020).

The colour of the Little Penguin's back, tail and upper head varies from a deep blue to a greyish-blue depending on the subspecies, the light, and if wet or dry. The chin and front are white. The eyes are also grey-blue with a darkish beak and pale pink feet with black soles. The juveniles are similar to the adults, whereas the chicks are greyish-brown and white.

The Little Penguin was also one of the three penguin species I had already seen before my husband died; we had seen colonies at Oamaru, the Otago Peninsula and Lyttelton Harbour, all on South Island, New Zealand. Although I only had very poor photographs, I did not envisage improving on these so did not specifically set out to find more as part of Mission Penguin. This is because the Little Penguin is the only penguin that is truly nocturnal on land, as the adults come to shore after dusk and leave before dawn. Photographing them is therefore extremely difficult, as flash would hurt their eyes and, even on moonlit nights, the low light leads to very grainy images.

Despite not targeting the Little Penguin, I seized on an unexpected opportunity to visit a colony one evening in Melbourne, Australia, while visiting friends. The Little Penguin is aptly named as it is the smallest of all the penguin species, and in Australia it is affectionately known as the Fairy Penguin. Other names include the Little Blue Penguin or even just the Blue Penguin, because of its bluish back. The colony we went to was at the St Kilda Breakwater, which was constructed to create a safe harbour for the sailing events in the 1956 Melbourne Olympics. The Little Penguins obviously decided that this was also a haven for them to raise their young, and a small colony emerged that is now a major tourist attraction in the bustling city harbour area. Although I only managed yet another poor photograph, I delighted in their antics as they came ashore, clambered over the rocks and waddled up the promenade.

BREEDING

Little Penguins are unusual in that, providing conditions are good, they can raise two, and very rarely three, clutches during a single breeding season. They nest in burrows, crevices, caves or under vegetation in small colonies. Breeding typically starts in late July but varies with location, and the duration depends upon the number of clutches. Two eggs are laid in each clutch, usually a few days apart, but hatch at around the same time and both chicks are then raised. They are brooded and provisioned alternately by both parents for the first three weeks. The chicks are then left alone in the burrow during daytime foraging trips, with the adults leaving at dawn and not returning until after sunset. The chicks typically fledge after 7–9 weeks.

Although there are several other colonies dotted around the southern coast of Australia, the majority of Little Penguins are found on small islands off the coast of Tasmania. While visiting this island I couldn't resist booking an evening boat tour from Strahan to Bonnet Island to view yet more of these beautiful little creatures as they returned from foraging. It was an even darker evening and so my photographs were even grainier, but it was still a delight to see them.

TOP RIGHT An individual Little Penguin at dusk on St Kilda pier, Melbourne, Australia.

RIGHT A pair of Little Penguins return safely to their burrow for the night on Bonnet Island, Tasmania, Australia.

PREVIOUS PAGES A Little Penguin on a rocky outcrop just off the Chatham Islands.

PREY AND PREDATORS

Little Penguins eat mainly small fish, including sprat, pilchards and anchovies, and also squid. The small size of the Little Penguin makes the adult birds, as well as the eggs and chicks, vulnerable to attack from a wide range of both aerial and land-based predators. These include large gulls, corvids, and introduced dogs, foxes, feral cats, stoats and ferrets, to name but a few. On Tasmania, quolls and Tasmanian Devils increase the death toll further, while at sea the penguins are predated by various sharks, sea lions and fur seals.

My photography breakthrough came unexpectedly as we approached the last New Zealand island group on the 'Birding Down Under' expedition, the Chatham Islands. We sailed past a small rocky outcrop in the south Pacific Ocean and there, sitting resting in broad daylight, were two Little Penguins. The ship passed close enough for me to capture a good photograph at last, which revealed the beautiful blue back and stunning pale silvery irises. I couldn't believe my luck, and finally had a photograph more befitting of Mission Penguin.

BELOW A Little Penguin near Chatham Island, New Zealand.

Seeing Little Penguins in Australia, New Zealand and the Chatham Islands may well prove to be more important than I had first realised. There are currently six recognised subspecies of Little Penguin and ongoing analysis and debate as to whether any of these should become separate species. Some scientists want to split the New Zealand and Australian groups into two species; others want to separate the current White-flippered subspecies into its own species, while some think this is simply a colour morph. As I had seen the various subspecies, Mission Penguin should thankfully be future-proofed.

ABOVE A White-flippered Little Penguin on the Banks peninsula, South Island, New Zealand.

LEFT A Little Penguin using a nest box on the Otago peninsula, South Island, New Zealand.

KEY THREATS

Many colonies of Little Penguins are close to human habitation and so are severely threatened by human disturbance, including loss of habitat for development, gillnet fishing and oil spills. They may also be killed and injured on roads and by watercraft. Climate change is also a major concern, as warming ocean temperatures threaten food supplies and increasing land temperatures can cause fatal overheating in adults and chicks. This is particularly a problem where nests are in vegetation rather than burrows or caves, and the provision of nest boxes has proved successful in some areas. Both Australia and New Zealand have designated the Little Penguin a fully protected species to help minimise direct human impacts, although the challenge of global warming will be rather more difficult to address.

Meet the
GREAT PENGUINS

Aptenodytes

As the group name suggests, the two great penguins are the largest of all the penguin species, with the 'greatest', the Emperor, reaching up to 1.3m (4.3ft) tall. Both are brightly coloured, with a yellowy-orange neck and ear patches that look stunning against the silvery-grey back and white front.

Other features the two species have in common are the relatively long breeding cycle and the fact that neither makes a nest. Instead, they lay a single egg which is carried with them on their feet and protected from the harsh weather by a brood patch.

King Penguin

Emperor Penguin

OPPOSITE   A King Penguin with its egg on its feet. The pink brood patch is just visible above the egg.

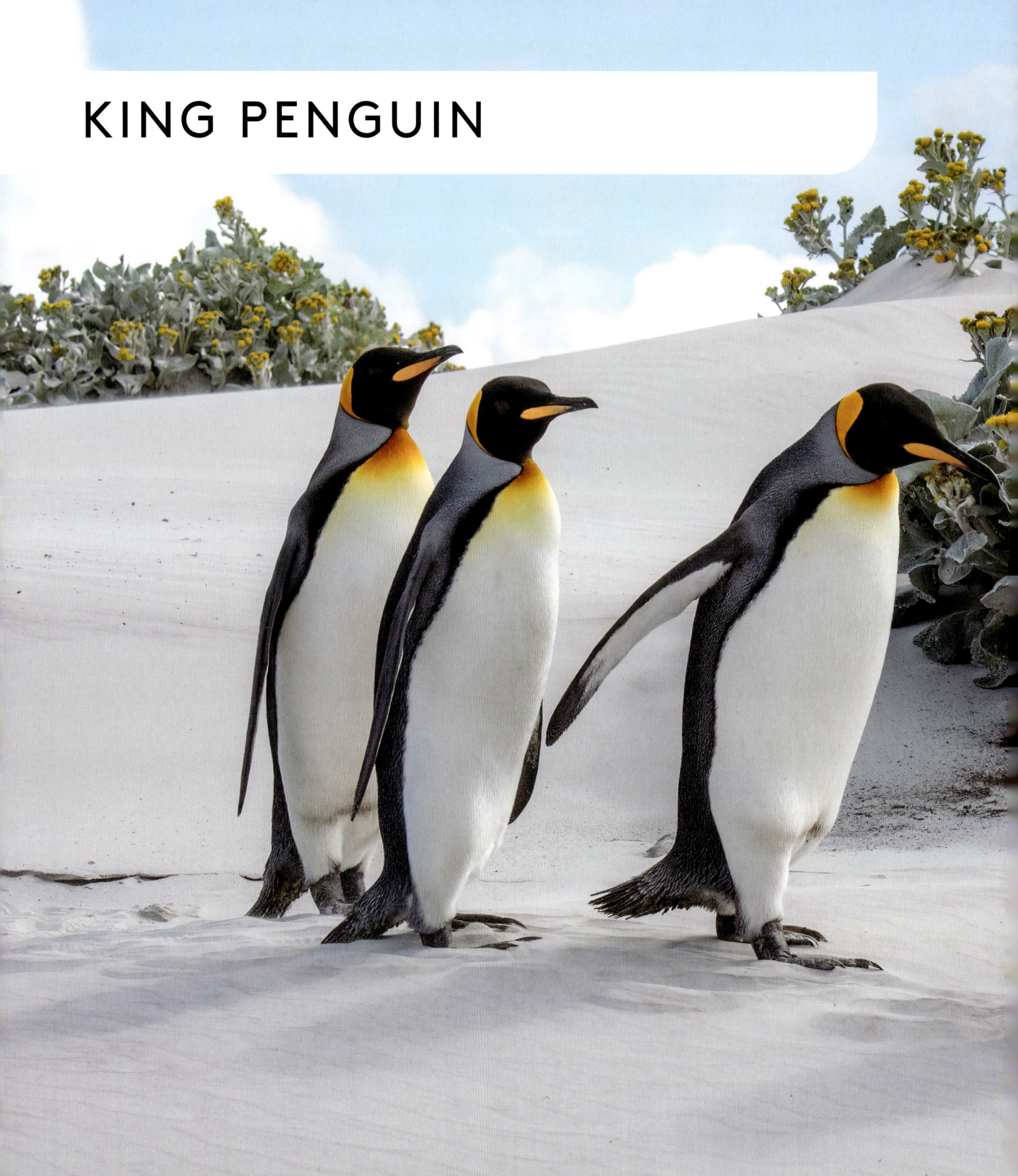

KING PENGUIN

KING PENGUIN *Aptenodytes patagonicus*	SIZE AND WEIGHT	POPULATION	CONSERVATION STATUS
	85–95cm (33.5–37.4in), 9.3–17.3kg (20.5–38.1lb) depending on time of year and gender.	Around 2.2 million mature individuals and increasing.	Least Concern (IUCN Red List 2020).

With a silvery grey-blue back, the King Penguin has black tail feathers, feet and head with orange ear patches. The deep orange and yellow continues down the neck, gradually fading into a white front. The long, thin beak is black with an orange-pink stripe below. Juveniles are smaller and duller. Chicks are born naked, developing a greyish-brown down before a thick brown woolly-looking coat.

When I saw a friend's photographs of her trip to South Georgia, I knew at once that was where I wanted to go to see the King Penguins. The colonies in her images stretched as far as the eye could see, with thousands of birds in each one. The destination for Mission Penguin was therefore set – well, at least in theory.

When the ship's engine issues took South Georgia off my Quark Expeditions itinerary I was absolutely devastated, as there are no King Penguin colonies on the Antarctic Peninsula and only an estimated 1,000–1,500 pairs on the Falkland Islands. However, the expedition crew and captain were brilliant and, knowing that several passengers, including me, were so disappointed not to have seen King Penguins on South Georgia, arranged an additional excursion before final disembarkation. This was to Volunteer Point on East Falkland, where most of the Falkland Island King Penguins breed.

When they announced their plan, I was thrilled. As the ship approached the island, I was out on deck frantically scanning the shoreline with my binoculars. The King Penguin is the second tallest of all the penguin species, which certainly aided my search.

PREVIOUS PAGES A procession of King Penguins crossing the sea cabbage-dotted dunes between the colony and the ocean, Volunteer Point, East Falkland.

RIGHT King Penguins on the beach at Volunteer Point, East Falkland, as the wind whips up the waves.

We finally drew close enough to make out the shapes of some King Penguins on the beach. These were still, however, too far away to photograph and I was desperate to get closer and land on the beach as planned. It was extremely windy and there were intense deliberations as to whether we could safely launch the Zodiacs to take us ashore. The weather conditions meant that we were approaching the safety limit and we were warned that if we chose to go it would be a bumpy and wet ride. They were not wrong, but the skill of the crew really shone through as we skimmed over the tops of the waves, even staying relatively dry. We landed safely on a beautiful pristine white sandy beach and were instantly greeted by groups of King Penguins going back and forth between the sea and colony. They were stunning, with the most vibrant face and neck markings of any of the penguin species. I didn't know where to point my camera next and took hundreds of photographs of individuals, pairs and groups in various poses. They were so comical, especially in their interactions, and I was entranced.

OPPPOSITE A King Penguin speckled with grains of sand that were constantly swirling in the strong winds.

BELOW Two taller male King Penguins stretching out their bodies and pointing skyward to attract the attention of the shorter female during courtship.

ABOVE The King Penguin is the brightest coloured of all the 18 species.

LEFT Squabbles often break out and flippers are used for slapping opponents.

OPPOSITE Penguins have dense plumage. The outer feathers overlap tightly, like fish scales, to protect a downy layer beneath.

Eventually I tore myself away and made my way over the dunes to the colony. Here even more treats were in store. One of the most amazing things about the King Penguin is its incredibly long breeding cycle, which lasts well over a year. This meant that the birds were all at different breeding stages in the colony at the same time. Within an hour I was able to observe and photograph pair-bonding behaviour, parents with an egg on their feet, tiny newly hatched chicks, older chicks in a crèche, huge chicks nearing fledging, and moulting adults. It was the complete cycle right in front of my eyes, and my Pingu-like flippers were flapping in excitement along with the penguins.

BELOW LEFT The King Penguin lays a single egg, which it incubates on its feet. To keep it warm, the egg sits against a bare brood patch and under a feathered fold of skin.

BELOW RIGHT The newly hatched chick is brooded on the feet and kept against the brood patch for warmth.

BREEDING

King Penguins breed in very large, dense, noisy colonies. However, they do not all lay their eggs at the same time, and have the longest breeding cycle of all the penguin species, lasting 14–16 months. As they only lay one egg per cycle this means that they usually raise two chicks every three years. Rather than build a nest, the single egg is held on top of the feet and incubated for about eight weeks against a bare brood patch covered by a fold of feathered skin. Both parents have this brood patch so share incubation by passing the egg between them. The chick is also brooded by both parents in shifts for the first 4–5 weeks, after which crèches form, enabling both parents to forage. At this stage the chicks are fed every 5–7 days, although this reduces over the winter months, and fledging occurs after 10–13 months.

BELOW Young King Penguin chick.

BELOW As the King Penguin chick grows, it develops brown downy feathers.

BELOW A King Penguin parent with newly hatched chick, Saunders Island, the Falklands.

OPPOSITE TOP After 4–5 weeks, the chicks are old enough to be left in a crèche while both parents go foraging for food.

OPPOSITE BOTTOM A King Penguin standing out from the crowd.

BELOW The King Penguin colony,
Volunteer Point, East Falkland.

I could have stayed there all day, but unfortunately the winds continued to pick up. Suddenly the cacophony of the colony was pierced by the sound of the ship's horn signalling an immediate return. The transfer back to the ship was extremely bumpy, but I was flying so high I didn't even notice the discomfort voiced by my fellow travellers. A fantastic day for Mission Penguin.

My second main sighting of King Penguins was on Macquarie Island in the south-west Pacific, while searching for the endemic Royal Penguin. I was particularly thrilled to see them there on two counts. Firstly, historically their numbers had been decimated when millions of King Penguins were killed for their oil, leaving just one colony of around 4,000 birds. Over the last 100 years their numbers have slowly increased and there are now over 100,000 breeding pairs at various colonies around the island. The second reason I was so pleased was that the King Penguins there belong to a different subspecies to the ones I had previously seen on the Falkland Islands – a double tick on this species for Mission Penguin.

It was also on Macquarie Island that I saw the largest penguin chick I had ever seen, and it was still being fed.

BELOW King Penguins at Sandy Bay, Macquarie Island.

ABOVE By the time King Penguin chicks finally fledge, they are huge. Early explorers thought that they were a separate species, which they named the 'Woolly Penguin' because of their thick brown 'woolly' coats.

RIGHT King Penguin chicks are fed a diet of regurgitated fish and squid.

PREY AND PREDATORS

King Penguins eat mainly fish. The eggs and chicks are predated by skuas and giant petrels while the adult birds are killed mainly by Orcas and Leopard Seals, or occasionally by South American Sea Lions and Antarctic Fur Seals.

The King Penguin subspecies on South Georgia is the same as that found on the Falkland Islands. There was therefore no justification for me to rebook a visit to South Georgia and I had accepted that I would never get to see the massive colonies I had originally envisaged. All this changed, however, four years later while searching for Moseley's Rockhopper Penguin on Tristan da Cunha, when our expedition ship called at South Georgia on the way. This turned out to be my favourite part of the trip, and to see over 100,000 King Penguins stretched out before me was far beyond anything I could have imagined. It was paradise and a dream come true.

ABOVE King Penguins on Salisbury Plain, South Georgia.

OPPOSITE TOP The colony of King Penguins at Fortuna Bay, South Georgia.

OPPOSITE BOTTOM King Penguins swimming alongside our boat.

BELOW About 95 per cent of the world population of Antarctic Fur Seals breed on South Georgia. The massive rebound in numbers following the end of their persecution has resulted in excessive competition with the penguins for both space and food.

ABOVE King Penguins, East Falkland.

OPPOSITE A King Penguin pair reinforcing their bond in light sleet. Although the markings are very similar, the male is slightly taller and heavier than the female.

KEY THREATS

The primary threats facing King Penguins are commercial fishing, oil pollution, disease and climate change, which affects the distribution of prey, especially for the more northern populations.

EMPEROR PENGUIN *Aptenodytes forsteri*	SIZE AND WEIGHT	POPULATION	CONSERVATION STATUS
	100–130cm (39.4–51.2in), 22–40kg (48.5–88.2lb) depending on time of year and gender.	Around 513,000 mature individuals and decreasing.	Near Threatened (IUCN Red List 2020).

The Emperor Penguin has a steel-grey back with black feet and a black tail. The white front turns to pale yellow towards the throat and up the sides of the neck with some deeper yellow-orange against the black head. The lower part of the black beak has an orange-lilac stripe. The juveniles are smaller and paler than the adults and the chicks have silvery-grey bodies with black and white heads.

The Emperor Penguin is the largest and probably the most iconic of all the penguin species. It therefore seemed fitting that it should also be the final species to complete Mission Penguin. Many adults, including myself, became aware of the incredible story of the life of Emperor Penguins through Luc Jacquet's award-winning 2005 documentary *March of the Penguins*. A year later the Emperor Penguin was also brought to the awareness of children through the computer-animated film *Happy Feet*. This film certainly reinforced my opinion that the chicks of the Emperor Penguin are the cutest of not only all the penguin species, but possibly all the birds in the world. I was, therefore, determined not just to see an adult Emperor Penguin but also to visit a colony and see these gorgeous chicks for myself.

The Emperor Penguin, along with the Adélie Penguin, is the most southerly breeding of all penguin species, so reaching a colony was certainly going to be a challenge.

The colony that is most often visited is on the south side of Snow Hill Island in the Weddell Sea. Attempts to access this colony occur in November, which is early summer in Antarctica. It requires an ice-strengthened ship, a helicopter and a trek on foot over the sea ice. The Oceanwide Expedition I found, 'In Search of the Emperor Penguin', clearly stressed that there was only a 30 per cent success rate in reaching the colony. In other words, seven out of every 10 trips failed due to either sea ice preventing the ship from reaching the island or bad weather grounding the helicopters. I was more than willing to take my chance and eagerly booked, only to be scuppered on my first attempt by the Covid-19 virus, which caused postponement until the following November.

Eventually, the departure date arrived. I was full of excitement, anticipation and also a slight dread of the first few days, which required crossing the notorious Drake Passage separating South America from Antarctica. It is

RIGHT Emperor Penguin chicks are incredibly cute.

OPPOSITE An Emperor Penguin and an Adélie Penguin on a distant ice floe in the Erebus and Terror Gulf, Antarctica.

PREVIOUS PAGES Emperor Penguins, Snow Hill Island, Antarctica.

known as either 'Drake Lake' or 'Drake Shake' depending on the weather, and I was mightily relieved that we experienced closer to the former. Before I left the UK, I had been researching the state of the sea ice in Antarctica and was growing increasingly confident that the ship would make it to the island. As we approached Antarctica, however, the winds started to build and for the rest of the expedition we spent most of the time battling hurricane-force winds. This severely impacted the proposed itinerary on the way to Snow Hill Island and I was beginning to

doubt that we would make it to the colony. I was therefore absolutely thrilled to see a single stray adult Emperor Penguin on a piece of floating ice, together with a lone Adélie Penguin. Mission Penguin was at least accomplished, and I celebrated with my fellow travellers and expedition crew that night. Despite this success I still went to bed dreaming of seeing a whole colony.

ABOVE Emperor Penguins travel constantly between the colony and the ocean.

OPPOSITE Two Emperor Penguins, with the colony in the far distance.

The next morning felt like a miracle when I awoke to blue sky, a calm sea and a gentle breeze just off Snow Hill Island. The helicopters were cleared to fly and the expedition to the colony was all systems go. I had to pinch myself to make sure I wasn't still dreaming. To make it fair to all, we were called to the launch pad by cabin number and the wait, although only a few hours, seemed an eternity. Finally it was my turn and we took off, flying over the snow-covered island. The scenery was spectacular, but my eyes were willing us onward, desperate to catch any glimpse of the distant colony. We landed about 1km (0.6 miles) away

and I began to follow the route carefully marked out for us by the expedition team. I initially tried to hurry over the snow-covered sea ice, but with each footstep sinking up to my knee I quickly learned that it was literally one step at a time. It also quickly became apparent that the journey to the colony was going to be almost as exciting as the destination, as adult Emperor Penguins lined our route. Small groups also waddled or tobogganed past us as they returned to sea to forage for their hungry chicks back at the colony.

The birds were so curious of us and so stunning to look at that I inevitably stopped and took lots of photographs before the pull of the colony drew me onward. As I continued stepping and sinking across the ice, I marvelled at the fact that the adult birds walk around 100 times further than this to reach the colony at the beginning of the breeding season.

PREY AND PREDATORS

Emperor Penguins eat mainly small fish, Antarctic Krill and squid. The eggs and chicks are predated by South Polar Skuas and Southern Giant Petrels, while the adult birds are killed mainly by Orcas and Leopard Seals.

BREEDING

Emperor Penguins breed on sea ice and in March walk between 50 and 120km (31–75 miles) to reach the colony. The males arrive first, closely followed by the females, and new pairs usually form each breeding season. The eggs are all laid at a similar time in the colony (May to early June) and the single egg is carefully passed to the male soon after laying. The females then leave the colony to forage, leaving the males to incubate the egg for around nine weeks during the extremely harsh winter. The egg is held on top of their feet against a bare brood patch and covered by a fold of feathered skin to keep it warm. The males also characteristically 'huddle', which involves constant shuffling in a circular motion to ensure they all take a turn at the coldest and warmest parts of the group. The females time their return to coincide with the hatching of the egg (mid-July to early August), at which point they swap duties and the males go foraging. They continue to alternate for 6–7 weeks, keeping the chick on their feet until it is large enough to join a crèche, enabling both parents to forage. The chicks fledge at about five months old (mid-December to early January), hopefully before the sea ice under the colony breaks up.

BELOW Emperor Penguins on the sea ice.

ABOVE An Emperor Penguin chick with both parents.

OPPOSITE The Emperor Penguin colony, Snow Hill Island, Antarctica.

When I finally reached the colony, I felt incredibly emotional and for a while struggled to see clearly through my camera viewfinder. I had made it, and the sight before me surpassed anything I had dreamed of. I continued smiling and clicking as I observed the chicks in crèches, parents returning and feeding their offspring, adults leaving, chicks playing. It was a magical experience that I will never forget.

TOP LEFT An Emperor Penguin chick and parent.

TOP RIGHT An Emperor Penguin family.

RIGHT The Snow Hill Island colony covers a large area.

TOP LEFT An Emperor Penguin chick practising its adult stance.

TOP RIGHT Emperor Penguin chicks are fed regurgitated fish, krill and squid.

LEFT An Emperor Penguin chick learning to 'toboggan'.

BELOW Emperor Penguin chicks form crèches while their parents are away foraging.

OPPOSITE Two Emperor Penguin chicks stretching out their flippers.

ABOVE Emperor Penguin portrait.

BELOW Emperor Penguin 'tobogganing' across the sea ice, which is faster and uses less energy than walking.

ABOVE The Emperor Penguin has stunning plumage.

BELOW An Emperor Penguin swimming across the melting sea ice as it leaves the colony.

ABOVE The sea ice beginning to melt around the colony.

OPPOSITE Two Emperor Penguins look back at the colony before continuing their trek to the Weddell Sea.

KEY THREATS

By far the greatest threat to the Emperor Penguin is global warming, which reduces the sea ice that is essential to their breeding success. Climate change also alters the distribution of prey, and commercial fishing is a further threat.

ABOVE The survival tents erected to protect us from the storm-force winds.

OPPOSITE King Penguins, South Georgia.

We were only allowed one hour at the colony as we then had to trek back to the helicopters, which were constantly shuttling passengers back to the ship. As I watched my allocated helicopter arrive, I noticed that it was unexpectedly still full. Apparently, a storm was brewing and gusts of 80 knots at the ship had prevented the helicopter from landing, forcing it to return to the colony. From that direction we could see a huge storm cloud approaching us and all helicopters were grounded. About 30 of us were now stranded on Snow Hill Island with the penguins. The expedition crew had already erected one survival tent at the site for emergencies and now quickly set about erecting a second. They expertly piled snow around the edges for extra strength and then ushered us all into them to sit out the storm. Although it was rather cramped and uncomfortable,

I felt safe, and the mood was one of excitement rather than fear. We were also entertained by the penguins, who had waddled over to see what was going on and were calling to us from just outside the tents. If we spoke 'penguin' I am sure that they would have been advising us to form a huddle.

After three hours the winds had eased sufficiently to relaunch the helicopters and the shuttling resumed. It was stopped again later and the survival procedure repeated for another hour, but I was already safely back on the ship and starting to look at my photographs. I could not stop smiling, and once everyone was safely back on board there were celebrations at dinner that evening. We had all achieved our expedition goal of seeing the Emperor Penguins, and for me it was the most wonderful and fitting end to Mission Penguin.

Mission Penguin accomplished!

AFTERWORD

After the most common question, 'why penguins?', I am then most often asked 'what next?'. In the early years of Mission Penguin, I quickly replied 'albatrosses', as these are also stunning birds to photograph and have an incredible life cycle. As Mission Penguin continued, however, my answer to this question changed markedly, for two key reasons. Firstly, Mission Penguin was my healing journey following the sudden death of my husband, and it worked. After a few years I discovered that I no longer needed to keep ultra-busy all the time but could relax again, confident in the knowledge that I was not going to fall apart. My new life is very different, but I am truly happy again and at peace. Secondly, over the decade it has taken me to complete Mission Penguin I have become increasingly aware of my own carbon footprint and the very real and present danger of global warming to our planet, including penguins.

According to the IUCN Red List of Threatened Species (2023), of the 18 species of penguin, five are listed as Endangered, four are Vulnerable and two are Near Threatened. Climate change is affecting penguins in many ways. Ocean warming leads to reduced quality and quantity of prey, causing chicks to starve.

In Antarctica there is reducing sea ice cover and earlier break-up of the sea ice. This has resulted in the loss of entire colonies of Emperor Penguin chicks as the sea ice literally melts beneath their feet before they have developed their waterproof plumage. Increased rainfall also affects other species, such as Chinstrap and Adélie Penguins, with chicks unable to stay sufficiently warm and dry so dying from hypothermia. For the more northern species, such as the African Penguin, warming land temperatures mean adults and chicks are dying from overheating.

Climate change is also leading to more dramatic El Niño effects. These reduce the uplift of nutrients from the bottom of the ocean, resulting in severe food shortages and subsequently the deaths of species such as the Galápagos, Humboldt and Magellanic Penguins. Increasing temperatures are also likely to facilitate the transmission of avian influenza (bird flu) with cases already reported from Galápagos to Antarctica.

Unless we act now, there is a very real possibility that several of the penguin species that helped me through my grief may become Critically Endangered or even extinct in the wild during my lifetime.

I implore everyone who has enjoyed this book to do whatever you can, however small, to help save our planet and make sure that these delightful creatures will still be around to help others like me.

OPPOSITE Erect-crested Penguin porpoising off the Bounty Islands.

MISSION PENGUIN EXPEDITION SUMMARY

EXPEDITION	PENGUIN SPECIES SIGHTINGS (First sighting of each species on Mission Penguin shown in bold type)	TRAVELLING COMPANIONS
Antarctic Peninsula and the Falkland Islands (Quark Expeditions; Audley Travel)	**Chinstrap, Adélie, Gentoo, Macaroni, Southern Rockhopper, Magellanic, King**	Lynn Love, Elizabeth Cole
Peru and the Galápagos (Audley Travel)	**Humboldt, Galápagos**	Sue Jenkins
Australia (Trailfinders)	**Little**	Sue Jenkins, Ellie Fletcher, Sally Bellingham
New Zealand and its Subantarctic Islands plus Macquarie Island (Audley Travel; Heritage Expeditions)	**Fiordland, Snares, Royal, Erect-crested, Yellow-eyed** Little, King, Gentoo, Southern Rockhopper	Sue Jenkins, Lynn Love, Patricia Fee
South Africa	**African**	Sue Jenkins, June Gregory, Clare Oldham
Tristan da Cunha via the Falkland Islands and South Georgia (Silversea)	**Moseley's (or Northern) Rockhopper** Macaroni, King, Gentoo, Southern Rockhopper, Magellanic, Chinstrap	Lynn Love, Sue Jenkins
Antarctica (Polar Routes; Oceanwide Expeditions)	**Emperor** Adélie, Chinstrap, Gentoo	Lynn Love
Falkland Islands (TravelLocal; Falkland Islands Holidays)	Adults and chicks of Gentoo, Magellanic, King, Southern Rockhopper, Macaroni	Hazel Prior

OPPOSITE Southern Rockhopper Penguins taking a shower on Saunders Island, the Falklands.

REFERENCES AND FURTHER READING

BirdLife International. 2024. IUCN Red List for Birds. datazone.birdlife.org

Clements, J. F. *et al.* 2023b. The eBird/Clements Checklist of Birds of the World: v2023. www.birds.cornell.edu/clementschecklist/download

De Roy, T. *et al.* 2013. *Penguins: Their World, Their Ways*. Christopher Helm (Bloomsbury), London.

Fretwell, P. T. *et al.* 2023. Record low 2022 Antarctic sea ice led to catastrophic breeding failure of emperor penguins. *Communications Earth & Environment* 4: 273.

IUCN. 2023. The IUCN Red List of Threatened Species. Version 2023–1. www.iucnredlist.org

Tyler, J. *et al.* 2020. Morphometric and genetic evidence for four species of gentoo penguin. *Ecology and Evolution* 10: 13836–13846.

ACKNOWLEDGEMENTS

I am indebted to all my travelling companions, travel agents, expedition companies and crews who helped me to achieve Mission Penguin. Huge thanks also to all those I met along the way who were inspired by my story and encouraged me to share it more widely. This book would not have been written without you.

Having decided to write the book, I am so grateful to the amazing wildlife photographer Sue Flood who encouraged me in my endeavour and put me in contact with Simon Bishop. Thank you so much Simon for believing in Mission Penguin and creating such a fabulous 'book layout and design' to share with potential publishers at the London book fair. When this was not possible due to Covid-19, you stepped in and emailed it to some of your publishing contacts, including Bloomsbury. I am so grateful, as without you I would not be writing these acknowledgements today.

So now to Bloomsbury – what a fabulous team you are. Thanks firstly to Jim Martin who picked up on the opportunity and then to my commissioning editor Alice Ward, who was always so positive and encouraging. Likewise, my editor Amy Hodkin, who also patiently dealt with all my queries and led me so expertly through the process. You are a star. Enormous thanks to Austin Taylor for his superb book design. Thanks also to Liz Drewitt, Lucy Beevor and Angie Hipkin for their work copy-editing, proofreading and indexing the book. It has been an absolute pleasure working with you all and I hope that the penguins made you smile too.

Special thanks to Lynn Love who got me travelling again after Ralph's death and whose visit to, and subsequent paintings of, Antarctica and South Georgia were a catalyst for Mission Penguin. Thank you for your ongoing friendship and some incredible adventures together.

Heartfelt thanks to my dearest friends, Monica Trenchard and Sue Jenkins, who picked me up after Ralph died and never put me down. Your love, friendship, support and encouragement know no bounds and I will be forever grateful. Sue also accompanied me on several of my expeditions, definitely going the extra mile, in more ways than one. What fun we have had and what memories we have made. Thank you from the bottom of my heart.

My wonderful brother, Jonathan Reece, has always been there for me and I know always will be. Likewise, Ralph's sister Rosemary Darrington, who is like a sister to me too. I am so blessed to have you both in my life and thank you for your immeasurable love and support.

Choral music was a central part of my life with Ralph and after his death I was determined to keep singing, however painful. I was welcomed into the hugely talented and super-friendly Taunton choir, In Ecclesia, who graciously tolerated my long penguin-related absences and supported me every step of the way. Thank you all for your warmth, encouragement, humour and of course the singing – you are fabulous.

Thanks also to Hazel Prior and Marian Remfry who, with Monica, are my fellow 'hummingbirds' in our a cappella quartet. Your friendship, music-making, hugs and giggles have definitely helped sustain me. And an extra mention for Hazel, who I managed to inspire (or indoctrinate) to love penguins almost as much as I do. Hazel included penguins in some of her wonderful novels and became a best-selling author, with one book even dedicated to me – something I will treasure for the rest of my life. Hazel, I will never forget our time together on the Falkland Islands. What a privilege and joy to be with you as you encountered penguins in the wild for the very first time.

Finally, for my dear departed husband Ralph, who showed me that the greatest gifts in life are to love and be loved. Love is stronger than death and your love continues to burn within me, supporting me daily in everything I do. Rest in peace my love.

OPPOSITE African Penguins preen each other to reinforce their life-long commitment.

PHOTOGRAPHER'S NOTES

As a young child I developed a passion for nature and am indebted to a junior schoolteacher who encouraged me to capture what I saw through photography. I began taking photographs at 10 years old and have never stopped. I find it the perfect way to relax and truly appreciate the beauty of the world around me.

I am self-taught and by far prefer taking the actual photograph over any subsequent editing, so keep this to an absolute minimum. I use two Nikon camera bodies and a variety of lenses. I tend to shoot on manual so select the aperture and shutter speed, but with automatic light sensitivity (ISO).

I encountered many photographic challenges during Mission Penguin. These included photographing from a small boat while being tossed around in winds and large ocean swells, as well as keeping my equipment warm in Antarctica and dry in the humidity of the Galápagos. Penguins, being black and white, also presented a particular challenge in terms of exposing the photographs correctly, keeping detail in the blacks while not burning out the whites. This was made even more difficult in bright sun which, although lovely, meant that perfect exposure could not always be achieved. I have still included these images in the book as they are part of my Mission Penguin story and I hope that you will forgive any imperfections.

Seeing wildlife is never guaranteed, even if you have travelled thousands of miles just to be there. I was so lucky that I managed to see every penguin species on the first attempt, although for some these were fleeting glimpses of just a handful of birds. This meant very few photographs and minimal, if any, choice for the book, hence some short chapters. For other species, however, my greatest challenge was selecting the images from the hundreds I had taken. I very much hope that you have enjoyed my choices.

More of my nature photography can be seen at www.f4inspirationalimages.co.uk

ABOVE Ursula Clare Franklin (Photograph kindly supplied by Hazel Prior).

OPPOSITE Adélie Penguin, Antarctica.

INDEX

Page numbers in *italic* refer to captions.